Living Landscapes

Hedges and Walls

First published in Great Britain in 2002
The National Trust Enterprises Ltd
36 Queen Anne's Gate
London
SW1H 9AS
www.nationaltrust.org.uk/bookshop

©The National Trust 2002

ISBN 0 7078 0322 5

Cataloguing in Publication Data is available from the British Library

Designed and art directed by Wildlife Art Ltd/www.wildlife-art.co.uk

All colour artwork by Dan Cole/Wildlife Art Ltd

Cover by Yellow Box Design with original artwork by Alison Lang

Typeset by SPAN Graphics Ltd, Crawley, West Sussex

Printed and bound in Italy by G. Canale & C.s.p.A

Living Landscapes

Hedges and Walls

Tom Williamson

THE NATIONAL TRUST

Contents

FACING PAGE A hedge-lined lane leads towards Pen-y-Fan in the Brecon Beacons.

Acknowledgements

Many people have helped with this book, providing advice, encouragement, information or inspiration. My thanks to colleagues, students and former students at the Centre of East Anglian Studies in the University of East Anglia, especially Kate Skipper, Susanna Wade Martins, Jenni Tanimoto, Jonathan Theobald and Richard Wilson; to all those involved in the Norfolk Hedge and Boundary Survey, particularly Lucy Whittle and Patsy Dallas; and to Gerry Barnes and Richard McMullen, who have over the years given much advice and information about hedges and hedging. I am also grateful to the National Trust conservation team in Cirencester, to Andy Wilson at the BTO, and to James Parry, who provided a mass of invaluable advice and information on natural history and ecology – this book could not have been written without him. Lastly, thanks as always to my family – Liz Bellamy, and Jessica and Matthew Williamson – for accompanying me on long walks in search of obscure forms of field boundary.

Foreword

Understanding landscapes is never easy. In a densely populated island like ours, the successive layers of man's presence are such that investigating the origins and history of a particular place is often a complicated and time-consuming process. It is, however, profoundly absorbing, and never more so than when the landscape concerned offers helpful pointers. Hedges, walls, banks and dykes are among the most revealing of these, acting as guides to past and present human activities and as refuges for a variety of fascinating wildlife.

At a time when we should welcome fresh approaches to landscape – and even need to decide what sort of countryside we want – it is essential to look at our environment as networks of interconnected elements. I am delighted that the National Trust has risen to this challenge with a new series of books on the social and natural history of different landscapes. The interface between agriculture, wildlife conservation and recreation in the countryside is a complex one, and can only be made clearer by the Trust's contribution.

Baroness Young of Old Scone
Chief Executive, The Environment Agency

Chapter one

❖

Introducing
field boundaries

INTRODUCING FIELD BOUNDARIES

The character of the British countryside is intimately related to the form and layout of field boundaries. The confused and intimate tangle of enclosures in Devon or Cornwall, bounded by great banks of stone and earth; the wide, rectilinear panoramas of Cambridgeshire, with fields defined by lines of flimsy hawthorn; the grim millstone grit walls of the Peak District: the lines of twisted, romantic Scots pines which edge the fields of East Anglia's Breckland; all define, perhaps more than anything else, the essential character of local landscapes.

This rich diversity was not created for our enjoyment. Hedges, walls and ditches were made for functional reasons: to mark the boundaries of property and, above all, to prevent livestock from straying, especially into areas planted with crops. Nevertheless, field boundaries have over time acquired an immense cultural significance. This is perhaps especially true of hedges. Although certainly not unique to Britain – they occur all over Europe, from western France to Crete – they are peculiarly identified with this country, and with England in particular. They are part of the national psyche. What would the paintings of Constable be without their tall and spreading hedgerows?

Field boundaries are part of our culture and our history. But they are also a vital part of our *natural* history. Hedges, even relatively recent and insubstantial ones, provide important refuges for wildlife and have influenced our very naming of nature, as in our use of terms like hedge sparrow and hedgehog. Other forms of boundary also have immense value as habitats. The huge hedgebanks of

Patchwork of ancient fields around the Sugar Loaf, Monmouthshire: hedges are as much a part of the Welsh landscape as they are of the English.

PREVIOUS PAGE An evening view along the Llŷn peninsula from Garn Fadryn hill fort, Gwynedd.

Pembrokeshire, for example, provide striking displays of wild flowers, with great cushions of thrift and stonecrop. Even the drystone walls of the bleak uplands in areas such as Malham in North Yorkshire often harbour rare ferns, lichens and mosses, as well as providing refuge for small mammals and nesting sites for birds.

The twentieth century was not kind to field boundaries. During the last fifty years many were removed, as relics of traditional farming systems rendered redundant by new technology, and torn out – like the hearts of so many of our historic towns – in a fashionable quest for neatness, efficiency and modernity. Others were retained but, being largely superfluous to the requirements of modern agriculture, were allowed to decay and deteriorate. Fortunately, at the start of a new millennium, the importance of field boundaries is becoming more widely recognised, and although problems of maintenance and management remain, we may be reasonably confident that the worst years of destruction have passed.

Dense, well-managed hedges – such as this fine example in Warwickshire – have become a rare sight in many parts of Britain.

A Wealth of Hedges

Hedges are the most common form of field boundary in Britain. Even in highland areas they were usually the preferred form of enclosure on lower, more sheltered ground. Most were deliberately planted, sometimes with a single species, but sometimes with several. A minority, however, may have developed in other ways. In some places they seem to have arisen spontaneously as plants were protected from grazing animals by a fence or wall: examples of such adventitious hedges can be seen developing beside some barbed wire fences and tumbled walls today. Others, it has been suggested, developed as lines of shrubby vegetation were deliberately preserved, and managed as boundaries, during the clearance of woodland in medieval times or earlier.

'Hedge' is itself a problematic term. Although we usually think of a hedge as a line of shrubs, managed to form a stock-proof barrier, even today the word has a wider range of meaning. Thus the substantial earth and stone banks which surround the fields in parts of Devon and Cornwall are called 'hedges' even though they are often only sparsely planted with vegetation. Indeed, in Anglo-Saxon times the word *gehægan* originally meant the creation of any form of barrier or fence, and early documents often refer to 'dead hedges', lines of brushwood staked to form

The ancient 'Penny Hedge' ceremony is still maintained at Whitby in Yorkshire.

a barrier. Some memory of this kind of hedge is maintained in the ceremony of the 'Penny Hedge', which is still kept up at Whitby in Yorkshire. In 1159 three knights hunting in Eskdale pursued a boar into a hermitage and, when the hermit attempted to save the animal, they attacked and mortally wounded him. On his deathbed he forgave them, but the Abbot of Whitby ordered that as penance they and their successors should forever, on the Eve of the Ascension, erect a hedge of stakes and branches on the foreshore at Whitby, sturdy enough to withstand the incoming tide. The 'penance hedge', or penny hedge, is still erected each year on the appointed day, at the appointed place. Definitions can be blurred in other ways. In many parts of south-eastern England, for example, there is no strict dividing line between a substantial hedge and a linear wood, Indeed, a short period of neglect can easily convert the former into the latter. The 'shaws' which are still such a feature of the Wealden countryside in Kent and Sussex reflect this seamless transition.

Walls and Other Boundaries

Hedgerows have received considerable attention from landscape historians and ecologists. Rather less has been paid to the other forms of field boundary, of which the most numerous are drystone walls. They are 'dry' in the sense that they are composed of stones which are not fixed together by mortar or any similar binding material. The stones are laid so skilfully that they are kept together by gravity alone. The majority of field walls are found in the upland areas of Britain: across much of Wales, in the Pennines and the North York Moors, on the high moorlands of Devon and Cornwall, and on the Mendips in Somerset. But they are also a feature of some areas of less elevated terrain where stone is readily available as a building material, notably the Cotswolds, the Isle of Purbeck in Dorset, and the 'heath' district to the south of Lincoln. In bleaker terrain the stones were sometimes simply picked from the surrounding ground: indeed, one function of such walls was to provide a repository for stones cleared from the surface of fields. But walling material was usually obtained from quarries, like those often seen on National Trust property in the Brecon Beacons.

Walls took up less space than hedges and required less regular maintenance. On the other hand, walls were more costly to create and, when repairs became necessary, they could prove expensive. Moreover, unlike hedges, walls had no additional functions, such as the production of wood: one eighteenth-century writer described a wall as 'an unproductive fence, and continually in need of

A panorama of stone walls in the Peak District. The light limestone walls are the most characteristic feature of the rolling landscape of the White Peak.

repair'. Their principal advantage was that they could be established in places so windswept, and on soils so poor, that hedges would not thrive. Walls provide shelter for livestock – especially ewes with young lambs – and are fireproof. This is a particular advantage where moors are managed by regular burning in order to encourage fresh heather growth for grouse and sheep.

As we shall see, forms of construction vary considerably from district to district. In part this is a consequence of the age of the walls in question, for methods of walling became more sophisticated (and, to some extent, more standardised) with time. But it also reflects the development of local traditions of walling which were adapted to the character of the local stone. Some rocks break easily into regular blocks; others into rather more irregular fragments; some are hard to break at all. Whatever their precise cause, variations in drystone walling contribute much to the character of local and regional landscapes.

The majority of fields in Britain are enclosed by walls or hedges. But a minority have rather different boundaries. In parts of western Britain, as already noted, various kinds of earth bank can be found, sometimes faced with stone, and with or without a line of vegetation growing along the top. And in areas of low-lying wetlands – coastal grazing marshes or peat fens – water-filled ditches serve not only to drain the land, but to prevent animals straying.

A magnificent display of thrift growing on a Cornish 'hedge' at Treknow Cliff, near Tintagel.

The value of field boundaries to small birds and mammals in turn attracts predators such as the sparrowhawk, often seen hunting along hedgerows.

It is not only the character of hedges or walls which is important. Equally vital are the patterns which they make in the landscape. From the shape of fields and the layout of their boundaries we can learn much about the development of farming and the history of local communities. Field boundaries, in all their variety and complexity, are fascinating historical documents.

Field Boundaries and Wildlife

Not only are field boundaries of immense interest to both archaeologists and historians, but they are also of considerable importance for wildlife. In intensively farmed arable districts in particular, where ancient grassland has been largely removed and where woods are often infrequent, hedges constitute one of the main refuges for many species of plant which were until recently common in the countryside. They are also often the only places where mammals and birds can make their homes and find sustenance in an otherwise bleak environment.

Some hedges are important because they contain species closely associated with ancient woodland, which are otherwise scarce in poorly wooded districts: trees like wild service and small-leaved lime and plants like wood anemone and primrose. But for the most part hedges contain fairly common scrub- and wood-edge species. They are not places in which to find rarities: they are instead refuges for everyday nature, and they are one of the places where nature is most frequently encountered. As hedges have grown rarer, or as their condition has deteriorated, nature and our experience of it have become impoverished.

As ecologists are well aware, the greatest diversity of species tends to be found at the boundaries between habitats, with the edges of woods being particularly rich in both plants and animals. Hedges form, in effect, a vast and extensive web of woodland edges and have for centuries served as 'corridors', whereby wildlife has been able to survive in, and move through, an increasingly cultivated and hostile landscape. Moreover, the distinctive structure of most hedges assures that they contain a particularly wide range of plants within a surprisingly small area, and this diversity sustains a correspondingly rich selection of insects, birds and mammals. In Chapter Four we shall look more closely at the range of wildlife species that hedgerows support.

But hedges do not have a monopoly on ecological interest. Most hedges, and especially those in areas of heavy soils, have an accompanying ditch and stand upon a bank. These features support additional communities of wildlife in their own right, thereby adding to the diversity provided by the field boundary itself. The grass verges that often run alongside hedgerows, especially those next to roads, are also a useful habitat for a variety of plants, small mammals and invertebrates.

As with hedges, the wildlife value of other types of field boundary, such as walls, banks and ditches, is greatest where the surrounding landscape is farmed most intensively. Here such features can offer welcome opportunities for plants, animals and birds to survive in what may otherwise be unpromising circumstances. For example, the potential value to wildlife of water-filled ditches in heavily drained areas like the Somerset Levels or the Fens of eastern England is quite clear. But perhaps less widely recognised is the conservation significance of walls, which in more gently rolling upland districts like the south Pennines can provide a substitute for scree, cliff and similar habitats. South-facing, well-drained drystone walls are especially important. However, in all districts walls provide shelter, and a measure of environmental diversity, for a wide range of flora and fauna. Chapter Five examines in more detail the different species that thrive in the habitats provided by walls, banks and dykes.

Field boundaries are central to our experience of the countryside. They represent, in an important and immediate sense, the place where nature and culture meet, and their character reflects the human, as well as the natural, history of each district. In the chapters that follow we will explore, and explain, their rich variety.

Butleigh Moor on the Somerset Levels. A ruler-straight 'rhine' (water-filled ditch) created by nineteenth-century enclosure and drainage.

Chapter two

❖

The earliest fields

Part of an extensive co-axial field system of Bronze Age date, surviving in the form of low, tumbled walls on Dartmoor.

THE EARLIEST FIELDS

Prehistoric and Roman Fields

Walls and hedges were a familiar feature of the British landscape by the time of the Roman invasion. The tumbled, denuded remains of walled fields of Bronze Age and Iron Age date can be found in many high, moorland areas. Sometimes these are arranged in irregular, haphazard patterns, but they can also exhibit a considerable degree of regularity, often taking the form of *co-axial* systems: that is, with long axes which run in one direction more persistently than in the other, producing a pattern of fields rather akin to slightly wavy brickwork. Arrangements such as this usually cover no more than a few hectares, but on Dartmoor the so-called 'reaves' form great blocks which can extend over more than 16 square miles. These striking field systems were laid out around 4,000 years ago, during the middle Bronze Age, when cultivation and settlement on the moor extended to a much higher altitude than today. Hedges were also common features of the British landscape by late prehistoric times. Excavations at the Iron Age site at Fisherwick on the Trent gravels, at Farmoor in Oxfordshire, and elsewhere have revealed twigs, almost certainly the cuttings from hedges, preserved in the waterlogged fill of the ditches and including specimens of hawthorn, blackthorn and hazel.

Upstanding traces of early field systems are largely restricted to tumbled, vestigial walls, or low earthwork banks, in what are now areas of marginal grazing land, especially on chalk downs and upland moors. Good examples can be visited on

PREVIOUS PAGE Fourteenth-century ploughing, from William Langland's *The Vision of Piers Plowman.*

18

the dramatic promontory of Brean Down in Somerset; on Fontmell and Melbury Downs in Dorset; on Pepperbox Hill in Wiltshire; on Box Hill in Sussex; and on Fyfield and Overton Downs near Avebury in Wiltshire. But aerial photography shows that prehistoric and Roman fields were once more widespread. Indeed, even on the heavier, damper soils which are unfavourable to this technique we can be confident that such fields once existed, for archaeological surveys have shown that by the Roman period farms and hamlets were found on almost all soils, and most would have been surrounded by at least some enclosures.

In a few places ancient fields seem to have survived in use, with only limited alteration, up to the present. The most striking examples are in Cornwall, on the Land's End peninsula and in West Penwith (particularly fine examples can be seen on National Trust land around Zennor). Here there is a magical landscape of small, irregularly shaped pasture fields, bounded by particularly massive earth and stone banks. Each has at its base a line of very large boulders, with smaller stones and earth above. These banks contain Bronze Age artefacts, and one incorporates an Iron Age *fogou* or underground chamber, archaeologists agreeing that the fields, and the banks that define them, came into existence during prehistoric times. They have survived with little or no subsequent change largely because of the difficulties involved in removing, or altering, such massive boundaries.

While such survivals are rare, it is possible that in other districts very early patterns of land allotment exist in more altered form. In some areas of lowland

A pattern of ancient fields, largely of prehistoric origin, still in use at Bosigran in West Penwith, Cornwall. The remains of 'courtyard houses' of probable Iron Age date can be seen in the centre and are clearly an integral part of the landscape.

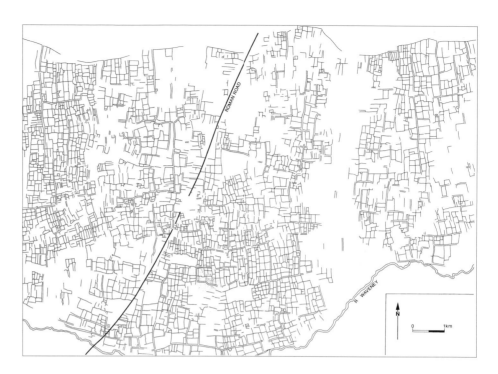

In south Norfolk the earliest layer of the modern landscape is a striking co-axial pattern defined by parallel north-south lanes, apparently slighted by the Roman Pye Road.

Britain the pattern of roads and boundaries appears to demonstrate a degree of large-scale planning, with organised landscapes extending over a far wider area than a single medieval manor or parish. Sometimes such apparently planned miles take the form of rough grids, like those which cover many square miles of the coastal plain and of the Dengie peninsula in south Essex, and which also occur – less distinctly – on the other side of the Thames in north Kent. In other cases they comprise co-axial, brickwork patterns, rather like hedged versions of the Dartmoor reaves. Such patterns are common on the dipslope of the Chiltern Hills in Hertfordshire and on the boulder clays of Norfolk and Suffolk. In some places Roman military roads appear to cut through such landscapes in a way analogous to a modern bypass or railway line, as at Yaxley in Suffolk or Dickleburgh in Norfolk, where an extensive co-axial field pattern is visibly slighted by a Roman road.

No one claims that all the boundaries in such landscapes are of prehistoric origin. Centuries of subsequent piecemeal change have removed much of the original detail. But in addition, in most cases the ancient elements seem only ever to have comprised a broad framework of roads and boundaries, and much of the field pattern was created by subsequent subdivision and infilling. In the case of the Dickleburgh 'system', for example, it is likely that the original landscape comprised a pattern of parallel drove ways, tracks along which cattle and pigs were taken to graze within areas of woodland and pasture, divided by comparatively few boundaries. This fairly open landscape formed the basic structure within which the ladder-like patterns of fields later developed in the course of the Middle Ages.

Medieval Fields

In most districts such early landscapes have not survived, for they were usually obliterated by later developments. In the fifth and sixth centuries the population of Britain appears to have fallen dramatically. The Anglo-Saxons, immigrants from the Continent who settled across much of lowland Britain at this time, entered a landscape in decline. Many areas were abandoned and the field systems fell into dereliction, to be cleared and divided anew as the population recovered from the eighth or ninth century. More importantly, across a broad swathe of England, running from Yorkshire through the Midlands to the south coast, the landscape appears to have undergone a radical revolution in later Saxon times. A pattern of scattered farms and hamlets, inherited in part from the Roman period, was swept away. Settlement became concentrated in a smaller number of nucleated villages and the surrounding landscape was re-planned. In these 'champion' districts, to use the phrase often employed by early topographic writers, there were few hedges outside the immediate vicinity of the village, and the arable land lay in 'open fields': extensive unhedged areas, in which the holdings of individual farmers were intermixed in a myriad of tiny strips.

In the areas to either side of this broad swathe, in south-east England and in the West Country, this revolution did not occur, or at least not to the same extent; and it is in these districts that most evidence for the survival of ancient fields in the landscape has been detected. These areas are often called 'ancient countrysides' by modern scholars, their landscapes characterised by winding lanes, ancient hedges, and a pattern of settlement which features large numbers of isolated farms and hamlets as well as, in some districts instead of, nucleated villages (see map on p.32). Hedged fields were already the dominant feature of these areas by the end of the seventeenth century, and in many of them enclosed land had predominated even in medieval times, although small areas of open field could often be found, as well as many greens and commons, exploited by the local community for grazing, fuel and much else. Early topographers often described these long-enclosed districts as 'woodland'. This was not primarily because they were densely wooded, although some were – like the Weald of Kent and Sussex, or the Chilterns – but referred instead to the large hedges and the numerous hedgerow trees which gave the landscape a 'bosky' appearance.

The 'champion' landscapes, of large villages and extensive open fields, reached their most developed form on the clay soils of the Midlands. Here the countryside was, by the thirteenth century, almost unrelieved arable, with most villages having at least 80% of their land under the plough. The strips or lands of each farmer – each normally between a quarter and half an acre in extent (up to 0.2 hectares) – were scattered, more or less evenly, throughout the territory of each township. The strips were grouped into bundles called furlongs, which formed the basic unit for cropping; these in turn were grouped into larger blocks called fields. There were usually three (but occasionally two or four) in each township, with at least one of these remaining fallow or uncropped each year. This provided crucial grazing for

the village livestock, which were herded or folded there, their dung returning to the soil the nitrogen and other nutrients that were constantly depleted by cropping. Farming activities were carried out, for the most part, on an individual basis, but the organisation of the farming year was subject to strong communal controls, usually dictated by the court of the principal manor of the village.

Rather different kinds of open 'champion' landscapes could be found on the light chalk soils of Wessex, the South Downs, the Chilterns, the Cotswolds and the Wolds of Lincolnshire and Yorkshire, as well as on the acid land of west Norfolk and west Suffolk. In these areas the ploughlands were more limited and the commons often very extensive, taking the form of tracts of heathland on the more acidic soils and of downland on the more calcareous. 'Sheep-corn husbandry' was the rule here, with large flocks of sheep grazed on the commons during the day and folded onto the arable at night. This ensured a constant flow of nutrients from one to the other, thereby allowing the poor leached soils characteristic of such light lands to be kept in good heart.

In the period after c.1500, and especially after 1700, these open landscapes gradually disappeared, leaving only a handful of much-altered remnants – of which the open-field village of Laxton in Nottinghamshire is the most famous. Yet in many places archaeological traces of open-field cultivation survive. On the heavy clay soils of the Midlands it was usual for each strip to be ploughed in the form of a low ridge and these are often preserved under pasture as prominent corrugations, usually around fifteen metres wide: a form of earthwork known to archaeologists as 'ridge and furrow'. The practice continued well into the eighteenth or even nineteenth centuries in many districts and was mainly intended to improve drainage. Agricultural intensification since the 1940s has removed many of these once common earthworks, but a number of striking examples survive. At the National Trust's Wimpole Park in Cambridgeshire an extensive landscape of relict open fields is preserved beneath the parkland that surrounds the mansion. Visitors today can inspect the ridge and furrow and see a variety of other early features that became fossilised beneath the parkland turf. These include hollow ways (old roads) and a circular mound constructed to support a medieval post mill.

There are other kinds of earthwork associated with medieval open-field farming. As population pressure built up during early medieval times, cultivation was often extended onto steep slopes, and characteristic terraces called strip lynchets were created. Some may have been the accidental side-effects of ploughing strips laid out along the contours (on slopes too steep to plough across the contours, the usual practice). Some, however, appear to have been intentionally excavated into the hillside, in order to increase the cultivated area and – perhaps – to reduce the problem of soil erosion. Such earthworks are particularly characteristic of chalk country, with notable examples at Cley Hill in Wiltshire and on Fontmell and Melbury Downs in Dorset. Perhaps the most striking example is at Glastonbury in

Well-preserved ridge and furrow in the National Trust's park at Wimpole, Cambridgeshire.

The striking terraces which help give Glastonbury Tor its strange, unnatural appearance are strip lynchets, created when the steep slopes were ploughed in the Middle Ages.

Somerset. Glastonbury Tor, the strange hill which dominates the town and abbey on this former island in the Somerset Levels, derives much of its unearthly appearance from the complex terraces that run round it. Mystics have interpreted these features in a variety of ways – for example, as evidence that the Tor itself is man-made, the remains of a mysterious 'initiation maze' – but in reality these are remarkable examples of strip lynchets and indeed the field immediately to the east of the Tor is still called 'Lynches'. The fact that medieval farmers were prepared to plough a slope as steep as this is testimony to the extent of land hunger in the many parts of England by the twelfth and thirteenth centuries.

Open fields were not entirely restricted to the central, 'champion' regions of medieval England. They were also widespread in the 'ancient countryside' areas of the south and west, although here they were often mixed up with hedged or walled fields and took a variety of different forms. Most were 'irregular' field systems, to use the jargon of historical geographers. The open fields were usually smaller and more numerous than in Midland districts (Gaddesden in Hertfordshire had twenty, for example): they were associated with the various small hamlets in a township, rather than with a single nucleated village. The holdings of individuals were thus clustered in the area closest to the farm, instead of being scattered evenly throughout the area of the township or parish, across two or three great, unhedged fields.

Open fields of various kinds were also wide spread in the rougher, rockier lands of the north and west of Britain. A particular precious survivor is preserved by the National Trust at Boscastle in Cornwall. The tenants crop the 42 curving strips or 'stitches' individually from Lady Day (25th March) to Michaelmas (29 September): then the whole area is thrown open for common grazing during the winter months. Where areas of low sheltered ground were extensive, nucleated villages developed, farming large open fields – the coastal lowlands of south Wales are a classic example. However, even in the uplands many areas of open arable could be found. In the mountains of Wales, for example, both the holdings of free kindreds, fragmented by repeated division between co-heirs, and those of bond communities, usually took the form of small open fields farmed from hamlets or loose agglomerations of farms. Some open fields were under permanent cultivation but sporadically-cultivated 'outfields' were common in most upland areas. Nevertheless, a much more important feature of these landscapes were the great tracks of upland moor, wild and uncultivated, that occupied the higher ground.

Yet these upland landscapes were not entirely unbounded in the Middle Ages. There were already significant lengths of drystone wall – generally of rough and often massive construction, made of material picked off the adjacent ground. There was usually a continuous wall – called a 'head dyke' in many northern districts – which separated the arable land on the lower slopes from the open grazing beyond. Some of these ancient stone walls, running for long distances along the upper ends or edges of vales and valleys, can still be seen, like the magnificent example (probably originally constructed in the thirteenth century) which surrounds the head of Great Langdale in Cumbria. Isolated seasonal grazing stations (called shielings in England and *hafodydd* in many parts of Wales) generally had walled enclosures around them, as did vaccaries, the specialised upland cattle farms owned by great lords and monastic houses. Moreover, by late medieval times walls had often been established more generally to define the areas grazed by particular communities.

In medieval Ireland we find similar themes. Here a variety of common fields also existed. Across much of the country *rundale* operated: groups of farmers, usually related and bearing the same surname, lived in loose agglomerations of farmsteads called clachans and had shares in a permanently cultivated infield, with outfield land and pasture beyond. Such arrangements survived into the nineteenth century – indeed, were sometimes created anew as late as this, as rampant population growth pushed settlement to the margins. Indeed, some

Ballyconagan, Rathlin Island, Northern Ireland: traces of cultivation ridges, associated with rundale farming, can be seen in the foreground.

rundale systems survived into the twentieth century around Lough Neagh and along the Antrim coast. In fertile areas more extensive open fields could be found, associated with relatively compact settlements, once again resembling the open fields of central areas of England. Fields were often cultivated in ridges, the earthwork traces of which can be seen in a number of places, as for example at Ballyconagan on Rathlin Island in County Antrim. As in Wales and parts of northern England, the exploitation of summer pastures from temporarily-occupied grazing stations was common in the more mountainous areas.

The landscapes of medieval Britain therefore displayed a great deal of regional variation. Much of this variety was directly related to the character of the natural environment – the distinction, in particular, between 'highland' and 'lowland' zones. Some may have been related to custom, tradition, and ethnicity. But some aspects remain mysterious, and in particular no one really understands the origins of the distinction between the 'champion' and 'woodland' areas of England. Some of the latter, such as the Weald of Kent and Sussex, were largely cleared and settled relatively late, in the eleventh or twelfth centuries. But others, such as Essex and Suffolk, were amongst the most populous parts of England in late Saxon times, to judge from the information contained in Domesday Book. Evidently, these districts did not experience the Saxon 'landscape revolution' which created the villages and open fields of the Midland belt, although why this should have been so remains unclear.

The Spread of Enclosure

Nobody knows for certain precisely how much land in medieval Britain lay within hedged or walled enclosures, and how much consisted of open fields and commons. What is clear is that the extent of enclosed land increased steadily from the late Middle Ages, so that by 1700 not much more than a third of the land of England and Wales still remained open, although in Northern Ireland enclosure generally came later.

'Enclosure' is a slippery word, with a number of meanings. It can mean the simple physical act of surrounding a parcel of ground with a wall, hedge or other barrier. But it also has legal implications. It was the process by which common grounds, shared by a number of individuals; or open fields, in which communal decisions determined land use and in which everyone had access to each parcel of ground at certain times of the year; were converted to landscapes held in severalty – that is, as full private property, which the owner could use as he pleased. Enclosure increased the importance of personal possession and decreased the importance of tradition and custom. The spread of hedges and walls across formerly open ground was thus intimately connected with broader patterns of social and economic change. Until the start of the fourteenth century, the population had been rising steadily in most areas of Britain. Most farms were small – thirty acres was considered a substantial holding – and mainly geared to mixed farming, with an emphasis on grain production. Some larger and more specialised holdings existed among the *demesnes* or home farms of manorial lords, but these only accounted for around a fifth of the land in England and Wales and most of these were primarily wheat and barley farms. With population pressing hard on resources, the main market was in cereals. But population growth was checked in the years around 1300 by a series of disastrous harvests, and by widespread disease among livestock. In 1348/9 the arrival of the Black Death led to a drastic fall in population. There were fewer people, and in relative terms wages were higher and rents lower. Farms became larger and, as living standards increased, farmers began to specialise in the production of a particular commodity or commodities. These developments were associated with wider changes in the character of the economy, and continued even when population growth picked up again in the sixteenth century. In some areas, particularly the light 'sheep-corn' lands, arable crops continued to be the most important product. But in others, especially districts of heavy soil, cattle farming or sheep grazing increased in importance, and by the sixteenth century a number of distinct farming regions had developed.

Until the eighteenth century most enclosure was associated with the expansion of pasture farming – it was hard to maintain large commercial herds in a landscape of open, intermingled strips. But not all areas well suited, by virtue of soils or climate, to livestock production or dairying could be enclosed with equal ease. Open fields disappeared earliest and most completely where they were small and 'irregular' in character: they lasted longest where they were most extensive and most complex – in the champion Midlands.

Enclosure could be brought about in a variety of ways, but it is helpful to make a broad distinction between piecemeal and general enclosure. The former involved the removal of open fields through a series of private agreements – sales and exchanges – which led to the gradual amalgamation, and subsequent fencing or hedging, of groups of contiguous strips. The latter, in contrast, involved the community of proprietors acting in concert to replan the landscape at a stroke, usually (although not invariably) enclosing all the open land in a township. The small and irregular open fields of 'woodland' and upland areas were particularly susceptible to piecemeal enclosure because holdings were less intermingled – each farmer had fewer neighbours in any one section of the fields.

Wardlow, Derbyshire. The distinctive pattern of gently curving drystone walls was created by the early piecemeal enclosure of open-field arable.

Piecemeal enclosure, because it involved gradual hedging or fencing along the margins of groups of strips, tended to preserve in simplified form the essential layout of the old landscape. Open-field strips seldom had absolutely straight boundaries. They were usually slightly sinuous in plan, characteristically taking the form of a shallow 'reversed S', caused by the way that the ploughman moved to the left with his team as he approached the headland at the end of the strip, in order to avoid too tight a turning circle. This sinuous pattern was fossilised by piecemeal enclosure, and in areas such as the southern Peak District whole tracts of countryside have field boundaries with a rippling, wave-like appearance.

In Wales, too, late medieval and early post-medieval times saw the progressive consolidation and enclosure of open-field arable. The continued operation of partible inheritance – the equal division of property between co-heirs – had reduced the economic viability of many holdings: the more successful and enterprising freeholders gradually bought out their neighbours, enclosing piecemeal as they did so. Even the more extensive open fields of the southern coastal lowlands had largely disappeared through this kind of enclosure by the middle of the eighteenth century, leaving the familiar pattern of parallel, curvilinear field boundaries, and in some places creating long, rather narrow fields – like those defined by the substantial earth banks ('Pembrokeshire banks') on National Trust property at Good Hope, to the east of Strumble Head in Pembrokeshire.

Piecemeal enclosure made much less headway in the champion areas of central England. When conversion to pasture occurred here in late medieval times it often took a more dramatic form. In the immediate wake of the Black Death, a number of settlements located on the Midland clays dwindled in size, partly due to their populations having been drastically reduced by the disease, but mainly because of the survivors leaving for land elsewhere that could offer better soil, more reliable water supplies or better access to markets. Faced with such incipient depopulation, landlords often hastened the process by buying up freehold land, evicting insecure copyhold tenants and re-letting the land in the form of large capital grazing farms. Many villages were deserted completely, and only their earthwork remains survive.

Enclosure of the Midland open fields continued right through the sixteenth and seventeenth centuries, even though the population began to grow again from the 1530s, and cereal prices started to recover. But it was now usually achieved 'by agreement': by the consent of the various proprietors who, having carried out the appropriate surveys, re-allotted intermixed arable land in consolidated holdings and divided the commons according to the rights formerly exercised over them. The kind of dramatic depopulating enclosure familiar from the period before 1530 thus came to an end, although enclosure was still often associated with some shrinkage of settlements, for grazing farms required fewer labourers than arable enterprises. For example, the parish of Great Linford in Buckinghamshire was enclosed in 1658, by the agreement of its principal proprietors. According to a

survey, of the eleven farms in the parish in 1714, only five seem to have included a farmhouse, and only three had any land in tilth – and even then, only a couple of hectares or so.

All these forms of 'general' enclosure, unlike those achieved piecemeal, tended to produce boundary patterns only loosely related, if at all, to earlier arrangements, with new hedges cutting boldly across older property boundaries, sometimes fossilised in the form of ridge and furrow. The earliest enclosures often produced very large fields. At Creslow in Buckinghamshire, enclosed in the early fifteenth century, the Great Field covered slightly more than 125 hectares. Later sixteenth- and seventeenth-century enclosures, in contrast, generally produced rather smaller fields. When Great Linford was enclosed in 1658 three great fields, along with small areas of common and meadow, were replaced by 53 new enclosures. Moreover, whereas the enclosures of the fifteenth century had generally created fields with fairly sinuous edges, those created in the seventeenth century were much more rectilinear in layout. And during this same period the large enclosures created in late medieval times were normally subdivided, producing a network of short, fairly straight hedges within the older, more sinuous boundaries.

The Limits to Enclosure

Enclosure in the period before c.1700 was largely concerned with the conversion of open fields to pasture: heaths, downs, moors and other kinds of common land survived almost intact. This was partly because they were largely immune to piecemeal enclosure – they were areas of shared use, rather than intermingled properties – but also due to the fact that most early enclosure was concerned with increasing livestock production and commons were, by definition, largely used for grazing. But some enclosure occurred, partly to improve the quality of the grazing. Many small greens and commons in 'woodland' areas disappeared in the course of the seventeenth century, and in the uplands manorial lords sometimes appropriated areas of grazing on the lower slopes, in the teeth of opposition from their tenants.

Furthermore, in parts of Northumberland and Durham in particular, landowners in the Restoration period and after sought to maximise their incomes by converting customary tenures to leases, and dividing the shielings and their grazing grounds from the lowland settlements to which they had traditionally been connected. On the National Trust's property around Housesteads on Hadrian's Wall, for example, research by Robert Woodside and James Crow has shown how some remote shielings evolved into permanently settled farmsteads in the late seventeenth century. Other new upland farms were established in the area on entirely new sites, and all were surrounded by new fields bounded by drystone walls. But tenants, too, were busy enclosing areas of fell and moor for their own use, a practice that was generally tolerated by manor courts. In addition, 'stinted pastures' or 'cow pastures' – enclosed areas which provided grazing for small groups of tenants – were often created on the lower slopes of fells. Their status

Only a few fragments of open fields have survived to the present day. These gently curving strips at the Vile at Rhossili in South Wales are known locally as 'landshares'. They are still unenclosed, but are now used mainly to grow vegetable crops.

lay somewhere between enclosed ground and common, but most soon became enclosed, for once separated from the open moor they were often divided between those holding stints, leading to the creation of walled fields appropriated to individual farms. Elsewhere – most notably in the southern Pennines in the later seventeenth century – large areas of common moorland were enclosed with the agreement of the principal proprietors. Nevertheless, although many inroads were made into the moors, vast areas remained completely open at the start of the eighteenth century.

In the uplands of Wales the situation was similar. Encroachments on moorland pastures were increasingly made in the sixteenth and seventeenth centuries, stimulated by the growing market for cattle and sheep in England. Here, too, summer grazing stations were steadily converted into permanently occupied farms. Moreover, as the population rose during the seventeenth and eighteenth centuries many illegal inroads were made into the moors by squatters. Nevertheless, vast tracts of the upland commons survived until enclosed by parliamentary acts which, as in England, were particularly concentrated during the Napoleonic Wars (1800-15).

The field patterns of England display a bewildering degree of variation. However, a useful distinction can be made between 'Predominantly Planned Countryside', in which the majority of boundaries post-date c.1650; 'Predominantly Ancient Countryside', in which most predate 1650; and the Highland Zone, in which ancient fields occur on the lower ground and the higher ground is occupied by open moor or fields created by late enclosure.

Extensive tracts of unenclosed commons also survived in low-lying wetland districts. Some wetlands, like the silt fens of Norfolk and Lincolnshire or the western parts of the Somerset Levels, had been divided and settled since early medieval times. But the great peatlands of England, and especially the southern Fens of East Anglia, remained as vast wet commons, exploited by communities living around their margins. Grazing occurred for some of the year but large areas were mown for hay or litter, cut for thatching materials, or dug for peat, while the reserves of fish and wildfowl were systematically exploited.

Highland Zone

Predominantly Ancient Countryside

Predominantly Planned Countryside

Wicken Fen in Cambridgeshire is one of the few surviving undrained portions of the Fenlands of East Anglia. The drainage windmill in the background originally stood elsewhere, but was re-erected in its current location by the National Trust.

As a result of the General Drainage Act of 1600 it became possible for large landowners to overrule local proprietors and suppress any common rights which obstructed the path of drainage schemes, the independent financiers of such schemes being rewarded with a share of the reclaimed land. In the middle decades of the seventeeth century a series of large scale drainage programmes was implemented in the Fens and a number of drains and new cuts created. As a result numerous blocks of drained land were allotted to the scheme's investors and to major landowners. Each block was bounded by straight dykes or ditches, rather than hedges: the watercourses acted as watery fences as well as drains. But large areas of the Fens remained as common land, with drainage conditions in many cases only marginally improved. Even in the main areas of drainage success was short-lived. Once water was removed, the surface of the peat shrank steadily and before long the land surface fell below that of the rivers. Windpumps were employed but most of the reclaimed land remained imperfectly drained. Nevertheless, the environmental damage was considerable, and extensive areas of rich fen were destroyed.

Elsewhere, the impact of enclosure on local wildlife varied according to the type of landscape. The changes in Midland districts, in particular, could be profound. Before enclosure these 'champion' parishes had few hedges or trees and little permanent pasture. Their ecology was dominated by arable weeds such as common poppy and corn marigold, and by the birds and animals that lived off these species and which were able to survive in what was a fairly bleak and open landscape. Enclosure, first and foremost, produced areas of permanent grassland in which, in time, a far wider range of flora and fauna came to be established. The range of niches was made particularly wide by the fossilised remains of the plough ridges which, within a limited area, ensured considerable variation in soil water content. Within a few decades ridge-and-furrow pastures came to be characterised by a large number of grass species and by a diverse collection of herbs, including such plants as betony, cowslip, yellow rattle, black knapweed and green-winged orchid. But equally important were the many thousands of miles of hedgerow established in this period, which provided shelter and breeding places for a wide range of wildlife.

A nineteenth-century illustration showing John Lilburne at the pillory in 1649. Hedges have long symbolised the concentration of ownership in the hands of the few, and during the English Civil War radicals like Lilburne and the Levellers violently opposed the enclosure of commons by the rich.

The social consequences of enclosure were generally less positive. The drainage and enclosure of the fen commons were opposed with particular vigour by those whose traditional lifestyle it threatened; indeed, the drainage works were often threatened by riots and vandalism. Many other enclosures also aroused fierce opposition, which on occasions erupted into local rebellions. Conversion to grass led to widespread unemployment, the loss of commons deprived the poorer members of society of their independence; everywhere existing forms of social and economic organisation, ruled by tradition and custom, were disrupted. We think of hedges as timeless and attractive features of the countryside, yet in earlier centuries they were the target for rioters. The governing classes were often worried about the social effects of open-field enclosure, in part because of the threat to social order it posed. From the early sixteenth to the late seventeenth centuries, attempts were made to limit and control the process. But the enclosure of common land was less strenuously opposed. 'The poor increase like flees and lice', noted one seventeenth-century commentator, 'and these vermin will eat us up if we do not enclose'. Most members of the ruling elite believed that commons encouraged idleness and vice, and attracted itinerants and other undesirables. In reality, the natural resources they offered provided labourers and small farmers with a measure of independence in an increasingly wage-dependent society.

The Uses of Field Boundaries

Hedges and walls were primarily intended to provide stock-proof barriers, but hedges in particular had a range of other functions. Like other practical features of the working countryside they could have an additional social and ideological significance: as statements of ownership, and as an expression of the fact that the land in question was enclosed and held in severalty, rather than being grazed in common, or cultivated under some kind of communal management. Cutting up an area of 'waste' into hedged or walled fields was almost a symbolic act, imposing order and ideas of private property across a threatening landscape.

However, in what was still, in effect, a peasant society, field boundaries also conferred a number of more direct and practical benefits. As reserves of woodland declined in the course of the medieval period, a higher and higher proportion of trees was concentrated within hedges. Most were managed as pollards – that is, their branches were repeatedly cut back to the trunk every ten to fifteen years in order to supply a regular crop of wood for fuel, tools and much else. Only a minority were left as timber trees. Of the trees mentioned in a survey of an estate at Thorndon in Suffolk in 1742, no less than 82% were pollards, 13% were classified as saplings and a mere 5% were timber. In early seventeenth-century Hertfordshire, to judge from a survey of crown estates carried out in 1605, pollards outnumbered timber trees by a ratio of nearly four to one. The relative proportions of pollards and timber in part reflected whether a farm was owner-occupied or tenanted and, if the latter, the extent of the control exercised by landlord over tenant. Traditionally, tenants had the right to take wood but not timber from the farm, and where the landlord was an absentee, the temptation

Pollarding was a way of producing a regular crop of wood for fuel, fencing and a host of other requirements. Trees were cut back to their trunk or bolling at a height of 2–3 metres, at intervals of ten to fifteen years, and the branches harvested.

to convert any young tree into a pollard was overwhelming. Pollards can live to a very great age, for the act of pollarding to some extent re-sets the tree's biological 'clock'. The restrictions imposed upon the development of the tree's crown delays the time when the demand for water and nutrients outstrips the capabilities of its root system. The majority of 'veteran trees' found in England and Wales are former pollards, and most of these stand, or once stood, in hedges.

In some districts of northern England walls incorporate pollarded trees, just as hedges often do in the south. Some particularly fine ash pollards, hosts to a range of rare lichens, can be seen growing in the walls of the National Trust's farm at Watendlath, just south of Derwentwater in the Lake District. In Swaledale in the north Yorkshire Pennines ancient elms, some probably with medieval origins, grow in walls now often ruined and tumbled.

An ancient pollarded ash, covered with lichen and growing in a drystone wall at Watendlath in the Lake District.

During the past three hundred years the overwhelming majority of farmland trees appear to have been oak, ash, and the various species of elm. In the Middle Ages black poplar, now a comparatively rare tree, also grew in many hedges. Oak was valued as a building timber, and to some extent for ship-building, but it was also widely pollarded in order to produce smaller wood. Elm is particularly versatile, resistant to decay and to splitting, and was widely used for hand tools, cart wheels and much else. Today, elm is seldom found as a hedgerow tree, owing to the ravages of Dutch elm disease, a virulent strain of which arrived in England in the mid-1960s. Ash, too, was widely used for tool handles and carts, and was particularly valued as firewood because of its welcome ability to burn even when green.

It was not only the trees within hedges that provided the farmer or landowner with a ready supply of wood. The hedges themselves also provided a rich store of wood, especially where they contained a good range of species like ash or elm, as well as thorn. In Somerset it has been suggested that the medieval pottery kilns at Donyatt were largely fuelled with material cut from nearby hedges rich in ash and oak. Arthur Young, writing in 1805, noted that in Essex the hedges were cut down at intervals of nine, ten or twelve years, and that fifty years before they had provided enough firewood to supply the requirements of the inhabitants of the county. In many districts, hedges were valued as much for fuel as for fences, and were planted accordingly. William Marshall thus described how in Hertfordshire in the late eighteenth century the local shortage of firewood had 'induced the farmers to fill the old hedges everywhere with oak, ash, sallow and with all sorts of plants more generally calculated for fuel than fences, and which would form no kind of fence under any management but their own'.

In the Middle Ages, and in some districts well into modern times, field boundaries could also be used to provide fodder – 'leafy hay' – for winter feed. Pollards were sometimes partly managed with this in mind but in addition some trees were regularly 'shredded'. A survey of the manor of Redgrave in Suffolk, made shortly after the Dissolution of the Monasteries in 1536, thus describes how on the 'seyd mannor and dyvers tenementes there ... be growing 1,100 okes of 60, 80 and 100 yeares growth part tymber parte usually cropped and shred'. Shredding involved shaving off the side branches of the tree so that it grew to resemble a kind of fuzzy toilet brush, producing an abundance of fresh, leafy growth. Oaks were the most commonly shredded species, although poplars and ash trees were also treated in this way.

Elm was also an important source of fodder and, in the Midlands especially, elm trees were widely planted in the hedges in and around medieval settlements. Where such villages were deserted and put down to grass in the fifteenth and sixteenth centuries the trees were often allowed to remain. Lines of ancient pollarded elms – now usually dead – are often found growing on the earthwork boundaries around such villages, and some may date back to the Middle Ages, although most are the direct descendants of such trees – if descendant is the right word for an identical genetic clone which has suckered from the same rootstock. Holly was also widely pollarded as a source of leafy hay, especially in districts lying in, and around, the Pennines. It is probable that in some areas hedges were themselves coppiced to provide fodder. Elm hedges were certainly planted in a number of areas of medieval England, and hedges containing a high proportion of holly are widely found, once again, in areas within and to the south of the Pennines, as well as in some parts of south-east England.

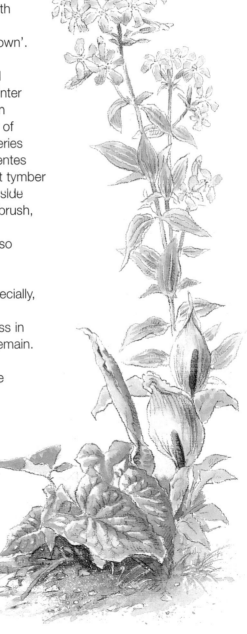

Historically hedgerows were regarded as a source of useful plants for use both in the home and in medicine. Soapwort (top) and cuckoo pint or lords-and-ladies (below) were used as washing and starching agents respectively.

The importance of hedges as a source of fuel and, perhaps, fodder, explains why in medieval and early modern times they were usually planted with a range of species. Indeed, in a host of ways hedges were a resource, not just a barrier. Fitzherbert's *Booke of Husbandrie* of 1598 thus urged the farmer to

'Gette thy quicksettes in the woode countrye and let theye be of whyte thorne and crabtree for they be beste, holye and hasell be good. And if thou dwelle in the playne country than mayste thou gete bothe asshe, oak and elm, for those wyll encrease moche woode in shorte space ...'

Thomas Tusser, writing in 1573, advocated the planting of elm, ash, crab, hazel, sallow and holly in hedges and, in praising the advantages of 'severall' or woodland countryside over champion, emphasised the fact that there was an abundance of fuel and fruit to be found in the hedges. Today people still visit the countryside in early autumn to collect the hedgerow harvest of blackberries: in earlier times a much wider range of hedge fruits was gathered. John Norden's *Surveyor's Dialogue* of 1610 commented on the abundance of fruit trees in the hedges of Devon, Gloucestershire, Worcestershire, Shropshire, Somerset and Kent, as well as of many parts of Wales, and lamented the fact that they were gradually disappearing from those in Middlesex and south Hertfordshire, as the modern generation failed to replace the trees that had grown old and died. It is noteworthy that many of the apple trees found in hedges today are not in fact the true crab (*Malus sylvestris* ssp. *Sylvestris*) but rather naturalised examples of the cultivated apple (*Malus sylvestris* ssp. *Mitis*). But crab apple trees are also abundant in many older hedges and in early modern times the crabs were fermented to produce a kind of cider called 'verjuice'. This, according to Tusser, had a range of uses:

'Be suer of vergis (a gallon at least)
So good for the kitchen, so needful for beast:
It helpeth thy cattle, so feeble and faint
If timely such cattle with it though acquaint.'

Nevertheless, the extent to which plants like bullace, crab or hazel (called the 'hedge nut' in Gerard's *Herball* of 1597) were actually planted in hedges with their fruit in mind is uncertain. The harvest of apples and nuts might have provided a welcome perk for the labourer, but the crops of nuts, in particular, would have been very exposed to the depredations of local wildlife. Some eighteenth-century writers disapproved of planting hazel in hedges specifically because of the damage which would be caused by trespassers in search of nuts! Also uncertain is the extent to which the other plants growing in, and at the base of, the hedge were systematically harvested in medieval and later times. Early herbalists sometimes describe uses for hedge plants which can never have been common. John Evelyn recorded making pickle from ash keys, but it is unlikely that this truly revolting concoction was ever widely eaten. Nevertheless, a number of useful

The bounty provided by hedgerows is particularly abundant in autumn, when a feast of hazelnuts, crab apples, sloes and other fruits becomes available.

plants can be found in hedges and in many districts
were largely restricted to them. Soapwort (locally
known as hedge pink) was often used for washing
and preparing cloth for fulling, before the introduction of
true soap; Good King Henry and jack-by-the-hedge were widely used as a
green vegetable, even in the nineteenth century in some rural districts; herb
robert ('robin run in the hedge', as it was known in Yorkshire) had a range of
medicinal uses; while ground ivy ('hedge maids' in Norfolk and Suffolk) was
employed as a medicine, and was the main source of bitter in brewing before the
widespread use of hops. Lords-and-ladies, a common plant of hedgerows, was
used as a source of starch for stiffening the cuffs and frills fashionable in Elizabethan
and Stuart times. The herbalist Gerard noted that the tubers produced 'the most
pure and white starch', but warned that it was 'most hurtfull for the hands of the
laundresse that hath the handling of it, for it choppeth, blistereth, and maketh the
hands ragged and rough'.

Early field boundaries thus fulfilled a range of functions, and did more than simply
define property or prevent livestock from straying. They served the varied needs of
a society which was still overwhelmingly rural, and in which a high proportion of
the population held and farmed at least some land. From the late seventeenth
century new forms of social and economic organisation developed apace. Hedges
and walls proliferated, but they now performed less diverse functions.

Chapter three

❖

The making of the modern fieldscape

THE MAKING OF THE MODERN FIELDSCAPE

The Agricultural Revolution and 'High Farming'

More than two-thirds of England and Wales were probably enclosed by 1700. But the following century and a half saw a massive increase in the numbers of walls, dykes and hedges and in many areas almost all existing boundaries are of eighteenth or nineteenth-century date. In Ireland, too, the majority of field boundaries were probably created in this period. Even in districts more anciently enclosed, numerous alterations were made to the field patterns in this period. The background to all these developments was the process of agricultural change often described as the 'agricultural revolution'.

The Acland family, who owned extensive estates in Devon and Somerset, were noted agricultural improvers during the eighteenth and nineteenth centuries. Francis Towne's painting of Holnicote emphasises the productivity, as well as the pastoral beauty, of the hedged fields.

The rapid rise in Britain's population – which had begun in the sixteenth century – tailed off after 1660, and demographic growth remained sluggish through the late seventeenth and early eighteenth centuries. From around 1750, however, the population began to grow rapidly once more. There had been some 7 million people in England and Wales in 1750: by 1850 this had grown to over 20 million. And with the development of large-scale industrialisation, more and more people were living in towns and cities, and working in mills and factories rather than on the land. Yet the nation was fed with the help of only minimal imports. Earlier centuries had seen improvements, often dramatic, in the practice of farming. But the late eighteenth and early nineteenth centuries saw a veritable revolution in agriculture, involving new crops and rotations, new farming methods, extensive schemes of land improvement, and the emergence of a new geography of farming.

PREVIOUS PAGE Intensive agriculture in Cambridgeshire; a solitary tree has survived the onslaught of modern farming.

Improvement continued in the middle decades of the nineteenth century, during the period of 'high farming', when mechanisation, new forms of drainage, manufactured animal feed and, above all, the widespread use of artificial fertilisers all pushed agricultural production to still higher levels.

This was also a period in which large estates came to dominate the landscape on an unprecedented scale, especially in arable farming areas, and wherever the soils were less fertile and the price of land comparatively low. Great estates played an important role in agricultural improvement, especially in Ireland, forcing the consolidation of holdings to create larger, more efficient farms, and further encouraging the adoption of new farming methods by their tenants. Above all, they were the prime movers in the enclosure of the remaining open fields, and in the expansion of cultivation at the expense of moorland, heaths and other 'waste' which was a central feature of this age of improvement.

Padbury in Buckinghamshire was enclosed by parliamentary act in 1795 and, like many parishes in the clay vales of the Midlands, was immediately laid to grass. The ruler-straight hedges, typical of those created by late planned enclosure, cut obliquely across the ridge and furrow: the new field pattern was quite unrelated to the earlier landscape.

In England and Wales, in the middle of the eighteenth century, parliamentary enclosure began to replace the various forms of enclosure 'by agreement'. A series of acts of parliament were passed which permitted the enclosure of particular townships or parishes and appointed commissioners to oversee the process. When all legal claims had been investigated and the existing pattern of properties and rights had been surveyed and recorded, an award was passed which created a new landscape of private property. This form of enclosure, unlike earlier methods, required not the agreement of the majority of landowners, but simply the assent of those holding a majority of the land – originally two-thirds, later three-quarters, of the acreage of a township. In the words of the historians the Hammonds, 'the suffrage was weighed, not counted'. Large and medium-sized landowners were thus able to overrule the opposition of small proprietors, and by the 1850s almost all open fields in England and Wales, and the majority of common land, had disappeared from the landscape.

Enclosure in the Lowlands

Parliamentary enclosures affected between a fifth and a quarter of the land of England and Wales and occurred in two great waves, the first peaking in the 1770s, the second around 1800, at the height of the Napoleonic Wars. The first wave largely affected the Midlands and represented a continuation of the steady drift into pasture which had continued fitfully here since the fifteenth century. The responses made by Midland vicars and rectors to a government enquiry of 1801 into the likely state of the year's harvest (the 'Crop Returns') frequently note the contraction of tillage in Midland districts. The vicar of Breedon on the Hill in Leicestershire typically described how:

> Within the last 30 years almost all the country north-west of Leicester
> to the extremity of the county has been enclosed: by which means
> the land is become in a higher state of cultivation than formerly; but
> on account of a great proportion of it being converted into pasturage
> much less food is produced than when it was more generally in tillage.

By the beginning of the nineteenth century pasture had come to dominate the landscape of the Midlands. Parishes which had long held out against enclosure now succumbed: ruler-straight hawthorn hedges divided their formerly open landscapes, often cutting obliquely across the ridges of former ploughlands, now preserved under grass. At the same time, the existing road network was often drastically transformed by the enclosure commissioners. Rights of way were curtailed, surviving roads were realigned, and a number of new roads usually laid out, some to provide access to new farms, established away from the village and conveniently located in the middle of their fields. Most enclosure roads are relatively straight and rather wide, with ample grass verges (usually 15–20 metres between the hedges): a distinctive feature of the Midland landscape. Wide roads allowed for the movement of stock, and also allowed travellers to find a dry route across a surface often churned up by passing sheep and cattle.

Enclosure and the spread of pasture on the Midland clays had other knock-on effects. In particular, it encouraged the rise of fox-hunting as a fashionable pastime. Until the end of the seventeenth century the hunting of foxes had been a low-key affair, which mainly involved surrounding the fox within its earth and setting dogs upon it. The classic form of the sport, involving a long chase across country, only developed in the course of the eighteenth century, and principally in Northamptonshire, Leicestershire, north Buckinghamshire, and Rutland – counties which have always held pride of place in the geography of hunting. In a landscape of enclosed pastures the gentry could ride fast in pursuit of their quarry, without damaging growing crops. Many landowners established coverts – small areas of gorse and other scrub, often interspersed with a few trees for ornamental purposes – where the fox could breed in safety, and where the huntsmen could be sure of finding their quarry. Over time these developed into the small woods in field corners which are such a characteristic feature of the landscape in parts of Leicestershire and Rutland especially. They now make excellent wildlife habitats in areas which are, for the most part, not otherwise well-endowed with woodland, and are in particular often frequented by various species of owl and woodpecker. The passion for hunting also had implications for local hedges. These were (and often still are) maintained with particular care, cut and laid with some regularity, in order to create a dense and low-growing wall of vegetation, an ideal obstacle to challenge the rider and horse.

Huntsmen in full pursuit of the fox. The grassing-down of the Midland shires was the main factor behind the rise of organised fox hunting here in the course of the eighteenth century.

Parliamentary enclosure was not everywhere concerned with the extension of grassland. In the traditional 'sheep-corn' lands of wold and down and heath it was associated with the expansion, and intensification, of arable husbandry. Turnips and clover began to be widely cultivated in the late seventeenth century, and were combined in regular rotations with wheat, barley and other grain crops. They provided good supplies of fodder and grazing, better than the thin sward of the downs and heaths, and it now made sense to plough up the old sheep walks and bring them into cultivation. At the same time, by growing clover and turnips in the fields, fallows could be eliminated. Instead of a landscape divided into permanent grazing and permanent arable – one half or one third of which lay fallow each year – entire townships could be given over to arable land, and yet at the same time more animals could be kept than ever before. More manure was thus produced, so that cereal yields were increased. But these changes were hard to achieve in an open landscape. The heaths and downs were, for the most part, common land, and could not easily be ploughed up without enclosure. Moreover, it could be difficult to introduce the new crops into open fields, for they disrupted the traditional patterns of folding and fallowing. Although turnips and clover were often cultivated in open fields, there is no doubt that their widespread adoption was greatly facilitated by enclosure.

Some enclosure in these light-land, 'sheep-corn' districts was achieved through general agreements, or through the simple expedient of large landowners buying up all the properties in a parish. But most was carried out by parliamentary acts, especially during the Napoleonic War years, when grain prices were high and

An unusual detail, painted on the enclosure award for Henlow in Bedfordshire, shows surveyors measuring land in the open fields prior to enclosure. A farmer looks on from the left. *c*.1798.

farmers and landowners were in an optimistic mood. Over the following decades the area of cultivation expanded inexorably. On the Lincolnshire Wolds, for example, by the 1850s it was said that 'the highest points are all in tillage, and the whole length of the Wolds is intersected by neat hawthorn hedges'. New sheep breeds were introduced, valued for their ability to fatten quickly on turnips and clover rather than to survive on the bleak commons and endure the nightly walk to the fold. And, as in the Midlands, enclosure was often accompanied by changes in the pattern of settlement. New farms were established out on the former heaths, wolds and downs, their remote locations often reflected in their names: Botany Bay, Quebec, New Zealand, Waterloo. They were equipped with large barns and – increasingly – regular rows of cattle sheds, enclosing a yard in which bullocks could be overwintered on turnips, their manure steadily piling up ready for removal to the fields. Today, the light lands of eastern England are quintessential landscapes of late enclosure. Whether created by parliamentary surveyors and commissioners, or by the agents of large landed estates, these are countrysides of straight-sided rectangular fields, usually rather larger than those created by eighteenth-century enclosure in the clayland Midlands: fields defined by species-poor hedges of hawthorn or, more rarely, sloe. The roads are often narrower than those laid out by enclosure commissioners in Midland areas and lack their wide verges.

The sheer scale of enclosure and reclamation in sheep-corn districts mesmerised contemporaries. The agricultural writer Arthur Young waxed lyrical about the changes taking place in north-west Norfolk during the second half of the eighteenth century:

> Instead of boundless wilds and uncultivated wastes inhabited by scarce anything but sheep, the country is all cut up into enclosures, cultivated in a most husbandlike manner, well peopled, and yielding an hundred times the produce that it did in its former state.

However, much grazing land did survive, especially on the chalklands of southern England. Moreover, the improvers often overreached themselves. Much of this poor, light land could not be profitably cultivated in the long term and was abandoned when prices fell back at the end of the Napoleonic Wars. William Cobbett, crossing the high chalk downs to the south-east of Winchester in 1823, described how:

> These hills are amongst the most barren of the Downs of England: yet a part of them was broken up during the rage for improvements … A man must be mad, or nearly mad, to sow wheat upon such a spot. However, a large part of what was enclosed has been thrown out again already, and the rest will be thrown out in a very few years. The Down itself was poor; what then must it be as corn-land! Think of the destruction which has taken place …

The New Hedges

The new hedges established by eighteenth- and nineteenth-century enclosure were, with some notable exceptions, very much a product of their age: one of growing standardisation and commercialisation, and in which ownership of land was being concentrated in fewer and fewer hands. Earlier hedges were planted to provide a range of products required by what was still, even in the seventeenth century, a largely peasant economy. Hedges planted in Georgian and Victorian times, in contrast, had more limited functions. Few farmers and landowners now looked to hedges as a major source of fruit, fuel or fodder. The increasing use of coal, in particular, ensured that they had less need for firewood, and few new pollards were established in hedges after the 1780s. Indeed, in many places, after the 1820s, few trees of any kind were planted there: most agricultural improvers advocated concentrating timber in plantations. Most hedges were composed entirely of hawthorn, a substantial minority of sloe, and a few were planted with both. These plants provided an impenetrable barrier but had few other benefits: one eighteenth-century writer shrewdly observed that hawthorn was a 'superfluous wood [which] will not repay the labour of planting and cutting, nor is it productive of any profitable fruit'. Other species were also occasionally used, sometimes plants which were alien to the region in question, or even exotics from abroad (see p.88).

Changing fashions in hedge planting reflected not only the more limited requirements of planters and the fact that many hedges were established where

Wharfedale in North Yorkshire. Straight drystone walls, created by nineteenth-century parliamentary enclosure, climb high up the slopes above Starbottom in the parish of Buckden.

areas of common 'waste' in particularly hostile environments were being enclosed, but also the development, in the eighteenth century, of large commercial nurseries. Small-scale local nurseries had existed since at least the seventeenth century, but large businesses able to supply standardised hedging plants in enormous quantities were a new development. By the 1770s most major towns could boast at least one such establishment. Their growth was stimulated in part by the mania for gardening shared by both the gentry and the prosperous middle classes, but a substantial proportion of their business was supplying trees for new plantations and 'quicks' for hedging. Estate accounts show that firms like Aram and Mackie of Norwich were able to provide many thousands of such plants in a single order. Not only did landowners thus have little need of hedges with a diverse range of plants: they no longer needed to acquire plants locally, from woods and existing hedges. Little wonder that most hedges created by eighteenth and nineteenth-century enclosure are still composed largely of lines of identical hawthorn. These were new kinds of hedge, for a new kind of society

Enclosure in the North and West

By the start of the eighteenth century relatively few common fields survived in the north and west of Britain, but vast areas of moorland remained open. Some enclosure of the lower, drier moors occurred in the course of the eighteenth century but it was only after c.1795 that attention really turned to the high, peat-covered plateaux and the bleak fells. A series of parliamentary acts, some dealing with many thousands of hectares of 'waste', was passed during and after the Napoleonic Wars. Others followed, albeit at a reduced rate, from the 1820s to the 1840s. Although many upland moors, especially in Wales and the Lake District, remained as common land, most had been at least technically enclosed by the middle of the nineteenth century.

On the lower moors enclosure was often followed by reclamation. Paring and burning, deep ploughing, draining, liming and reseeding were all employed to convert the heather and rough grass of the Mendip Hills, and much of the North York Moors, into arable land or improved pastures. But in the early nineteenth century attempts were also made to reclaim the higher, more windswept moors, and even to farm them as arable land. The gentleman farmer John Nicholson obtained a compact allotment of moor, rising to 385 metres above sea level, following the enclosure of Whinfell in Cumbria in 1826. This he surrounded and subdivided with drystone walls, and built a new farmstead: stones were cleared from the new fields, the gorse burnt and the land harrowed and sown. But here, as in many places, reclamation was destined to fail. The farm was abandoned by 1850 and most of the land reverted to rough grazing. Such short-term attempts at improvement in the early nineteenth century were widespread, and are often indicated today by the earthworks known as 'narrow rig', a diminutive form of ridge and furrow (with the furrows spaced at c.4m or less) principally associated with ploughing of late eighteenth- or early nineteenth-century date. Good examples can be seen on National Trust land near Housesteads fort on Hadrian's Wall.

However, enclosure of the higher moors was not always, or even usually, followed by reclamation. Where the blanket bog lay thick, or where on higher ground thin soil gave way to bare rock, improvement was impossible, and enclosure served simply to divide the moors into huge blocks, defined by stone walls, rather than to subdivide it into cultivated fields. Often the highest moors continued to be used in common by groups of farmers, as they are to this day, although the 'stints' – the number of animals they could turn out on to the high pastures – were now more carefully regulated, and the moors of each parish or township were clearly demarcated, by stone walls, from those of the next.

Major landowners often had no intention of converting the moors to agriculture and enclosure had a rather different purpose. Upland moors were often rich in minerals, and in this age of industrial expansion many landowners were keen to enclose simply to ensure the safe enjoyment of their mineral rights. Moreover, their enthusiasm for shooting red grouse increased in the course of the eighteenth and nineteenth centuries and enclosure increased the opportunities for game preservation at a time when improvements in gun technology encouraged more intensive and competitive forms of shooting. Instead of stalking the birds and shooting at them as they flew away, the grouse were now driven towards a line of sportsmen, concealed in shelters called butts. Parliamentary enclosure may actually have been instrumental in the development of this new mode of shooting, for the proliferation of stone walls on the moors made it more difficult to follow grouse over long distances while, conversely, the walls often provided the first rudimentary butts for the sportsmen.

Old grouse-shooting butts are a common sight on heather moors in northern England and were often constructed of stones taken from redundant or derelict walls.

The new areas of land allotted by an enclosure had to be walled with great rapidity, usually within twelve months of an award, and this provided considerable employment for local or itinerant wallers. Indeed, the late eighteenth and early nineteenth centuries have justly been described as 'the golden age of the professional waller'. Walling was a labour-intensive, slow business if modern experience is anything to go by: a waller working on his own will average around 5–5.5 metres per day. The work was carried out, for the most part, by professional gangs, sometimes (as in parts of west Yorkshire) composed of men whose main occupation was mining. In many moorland districts the walls created at this time are awe-inspiring, for the abstract grid of private possession was often imposed upon the landscape by enclosure commissioners with little thought to the topography, so that ruler-straight walls climb impossible gradients, or cut obliquely across steep-sided valleys. Sometimes (especially in the Yorkshire Pennines) the moorland enclosures take the form of long strips, each covering twenty hectares or more, and the ruler-straight, parallel walls reach out from the moor's edge towards the high watersheds running between the vales.

These new walls were not only patently artificial impositions on the natural topography. They were also remarkably standardised in their construction, for all the walls within a particular township or group of townships were usually built to the same basic specification, laid down in the enclosure award. At Grassington in Yorkshire, for example, the commissioners decreed that the new fields should be enclosed

> ... by good stone Walls, in all places made 34 Inches broad in the Bottom and 6 Feet high, under a Stone not exceeding 4 Inches in thickness, which shall be laid upon, and cover the Tops of the Walls in every Part, that there shall be laid in a Workman-like Manner 21 good Throughs in every Rood of Fence, and the first 12 to be laid at a height of 2 Feet broad, and the second 9 to be laid on at the height of 4 Feet from the Ground, and the Wall Batter to decrease gradually from the Bottom to the Tops which shall not be less anywhere than 16 Inches broad under the uppermost stone ...

The eighteenth and nineteenth centuries thus saw profound changes to the landscapes of upland Britain. But more dramatic still were those which occurred in the north of Ireland in this period. Traditional Irish forms of agriculture, based on rundale and other common-field systems, largely survived under the new English rulers established by Tudor and Stuart conquest. Only in Ulster were there significant changes. Here large-scale colonisation by English and Scottish farmers occurred after 1610, and although the Articles of Plantation insisted that colonists lived in compact villages for mutual protection, many occupied isolated farms within enclosed fields. Nevertheless, the main period of enclosure was from the middle of the eighteenth century. At first the initiative was taken by individual farmers, keen to expand the production of livestock for the English market. But

Hedges and walls

This map shows the complicated mosaic of farming regions which had developed in England during the sixteenth and seventeenth centuries (see above, pp. 27–8). In the course of the 'agricultural revolution' period a much simpler pattern emerged. As the Midland clays were laid to grass, the Fenlands drained and the ancient pastures of East Anglia put to the plough, England became divided – as it remains today – into a largely pastoral west and a mainly arable south and east (after Joan Thirsk).

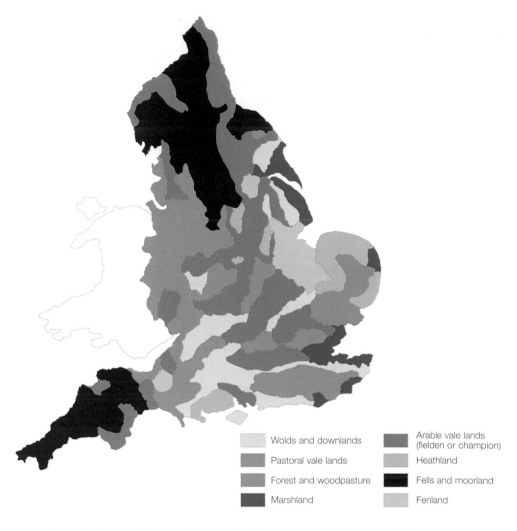

Wolds and downlands

Pastoral vale lands

Forest and woodpasture

Marshland

Arable vale lands (fielden or champion)

Heathland

Fells and moorland

Fenland

more important were the activities of landlords, eager to increase their rents at a time when the market for grain was also increasing. Farms were consolidated, rundale and other forms of common farming abolished, and often areas of long-enclosed fields replanned. Across much of the country neat networks of square fields were laid out, arranged in compact holdings. Such rectilinear field patterns, very similar to those created by parliamentary enclosure in England, were also adopted in the latter part of the nineteenth century when the government assumed responsibility for consolidation and improvement. In some places, however – especially in more mountainous terrain – distinctive ladder-like patterns were produced, with long axes running from the valleys on to the higher ground: each strip, divided laterally into small fields, represented a different farm, and farmsteads were often arranged in a line, along a road, at the end of the strips. The new pattern of settlement thus perpetuated elements of the old clachan, while the pattern of land allotment preserved some of the traditions of rundale, with an equal distribution of good and bad land.

The Expansion of Arable

The intensification and expansion of arable farming which accompanied enclosure on the light lands of the south and east of England, and the steady expansion of pasture in the Midlands, were part of a wider shift in the geography of farming. In the sixteenth and seventeenth centuries Britain had comprised a complex mosaic of farming regions, with some arable districts in the west and many pasture-farming areas in the east. In the period after c.1750 this gradually changed, and when James Caird came to produce a map of agriculture in 1852 he posited a much simpler pattern, one which – with important modifications – is with us still: arable farming was now largely concentrated in the east of the country, and grass in the west.

Dramatic changes occurred on the heavy clay soils of south-east England and East Anglia. Where arable farming already formed a prominent part of the local economy, as in much of Hertfordshire, new methods were adopted and production intensified. Where, as on the claylands of Suffolk and Norfolk, pasture was still predominant, it now fell to the plough. Much of this land required better drainage, especially if turnips were to be cultivated. In the eighteenth century farmers began to use 'bush' drains – trenches cut across fields which were filled with brushwood cut from hedges and pollards, capped with straw or furze, and then back-filled with soil. These were replaced from the middle years of the nineteenth century by pipes and tiles made of earthenware, which although more expensive usually lasted longer.

Parliamentary enclosure had some effect on these increasingly arable districts of the south and east. Many large commons, and small greens, were removed by enclosure acts, especially during the Napoleonic Wars. But these were, by and large, landscapes of ancient enclosure, with small irregular fields bounded by tall hedges, stuffed with pollards and timber. Sir John Parnell, writing in 1769, thus commented that he knew 'no part of England more beautiful in its stile than Hertfordshire. Thru'out the Oak and Elm hedgerows appear rather the work of Nature than Plantations, generally Extending 30 or 40 feet Broad growing Irregularly in these stripes and giving the fields the air of being Reclaim'd from a general tract of woodland'.

As the character of farming changed, so too did this bosky landscape. Small, oddly-shaped fields were inconvenient for ploughing, while dense and spreading pollards shaded out the crop and robbed the soil of nutrients, and wide outgrown hedges took up large areas of potentially productive land. So hedges were often drastically cut back, and the timber trees and pollards growing within them were systematically thinned. Sometimes, indeed, hedges were grubbed out altogether, in order to create larger parcels more suitable for ploughing. The rector of Rayne in Essex typically observed in the 1790s how on one farm in his parish 'the fields were over-run with wood', but 'since Mr Rolfe has purchased them, he has improved them by grubbing up the hedgerows and laying the fields together'.

Not all the wetlands in the south and east of England were ploughed during the agricultural revolution. Many areas of coastal marsh survived as pasture, as on Romney Marsh, although even here much land was in tilth by 1800. Fairfield Church, Walland Marsh, Kent, in 1904.

A government enquiry into the state of the nation's timber supplies was initiated in 1791 and included the question: 'Whether the Growth of Oak Timber in Hedge Rows is generally encouraged, or whether the grubbing up of Hedge Rows for the enlarging of fields, and improving Arable Ground, is become common in those Counties?' The answers received make it clear that the grubbing-out of hedges and removal of hedgerow trees, were widespread in the anciently enclosed districts of the south and east. Both were said to be 'frequent', 'becoming common', or the 'general practice'. One respondent typically noted that 'The county of Hertfordshire consists chiefly of Land in Tillage, and by clearing the Hedges of all kinds of Trees they admit of plowing to the utmost Bounds of their Land'. Thomas Preston from Suffolk had particularly firm views on the subject. He quite baldly stated that 'Much Timber and the Improvement of Arable Land are incompatible. Arable land in Suffolk is improved, and therefore timber is lessened'. The tidying up of 'ancient countryside' was also a feature of some western districts, especially Herefordshire, and in many areas it seems to have continued – perhaps with increasing intensity – into the 'high farming' period of Victoria's reign, encouraged by the adoption of traction engines for ploughing, cumbersome technology which required large rectangular fields.

The expansion and intensification of cereal growing on the claylands of the South East and East Anglia was accompanied by more dramatic changes in the Fenland of Norfolk, Lincolnshire and Cambridgeshire. In the decades either side of 1800 a great wave of parliamentary enclosure removed almost all the remaining fen commons, covering the landscape with a mesh of ruler-straight drainage dykes. There were major changes to the main arterial drainage channels and, in 1817, the first steam pumping engine was erected, at Sutton St Edmund in Lincolnshire. By the 1840s steam drainage had spread throughout the fens, and the old windpumps were steadily abandoned. By the 1850s the last remaining areas of undrained mire and fen were disappearing. As drainage improved the fens were steadily ploughed and used to grow potatoes, oats and increasing quantities of wheat. Indeed, by 1836 one witness to a government enquiry lamented how the price of wheat in England had been adversely effected by the 'immense tracts of land brought into cultivation in the fens of Lincolnshire, Cambridgeshire and Norfolk'. By the middle of the nineteenth century the East Anglian fens had assumed their modern appearance: a level panorama of unrelieved ploughland, stretching to the far horizon.

Cost and Benefits

The agricultural revolution and 'high farming' most certainly succeeded in feeding the rapidly growing population of the world's first industrial nation. But there was a price to be paid for all this additional food. The destruction of thousands upon thousands of hectares of ancient, semi-natural habitats – acid heath, chalk heath, downland and fen – was an ecological disaster on an awesome scale. Plant communities once common became rare and the archaeological remains of ancient times were flattened in a single-minded search for profit. The nineteenth-century botanist Claridge Druce was told how soldier and monkey orchids had been 'tolerably plentiful' on the chalklands around Whitchurch in Oxfordshire until the 1830s, when the steep slopes had been pared and burnt, 'roasting alive' both types of plant.

Fauna, too, was adversely affected – both by enclosure itself and by the agricultural intensification which often accompanied it. On the heathlands of the East Anglian Brecks the great bustard declined rapidly until by the 1840s only a few isolated female birds remained, the population effectively doomed. The bustard had made its nest in the fields of rye, sown broadcast, which was the traditional winter

Today we rightly regard hedges as being beneficial to wildlife, but ironically it was their very creation on the Brecks of East Anglia which spelled disaster for the last British population of great bustard.

crop on these poor soils. Enclosure, soil improvement and the adoption of the new rotations allowed the widespread substitution of wheat, sown in drills and weeded by gangs of labourers: few nests survived. But in addition, the bustard was an open-country bird, and the ornithologist Stevenson observed how Breckland was now increasingly subdivided by hawthorn hedges, rows of pines and tree belts,

> ... not only entirely changing its aspect but rendering it entirely unsuitable to the wary habits of the bustard, which soon learned to become as jealous as any strategist of what might afford an enemy harbour.

Of course, the effects of enclosure were not entirely negative. Where enclosure was associated with the expansion of permanent pasture – as across much of the Midlands – there must have been real benefits for wildlife. And the hedges of sloe and hawthorn offered some compensation for what was lost when ancient heaths and commons were ploughed; what proved a disaster for open-country birds like the bustard might be a boon for species such as wren or blackbird, which flourish in a more scrubby environment. In a similar way, while the hare probably suffered from the agricultural changes of the eighteenth and nineteenth centuries the rabbit certainly benefited. Before 1700 it was only really common in areas of light soil, often close to the warrens to which had originally been introduced in the Middle Ages. The spread of hedges allowed it to colonise more widely, especially in the Midlands, while the widespread cultivation of turnips provided welcome sustenance during the winter months.

There were thus winners and losers from enclosure and from the various changes that generally accompanied it. But in most districts losers were in the majority. Charles Barrington reported how until the turn of the nineteenth century much of the chalkland in the south of Cambridgeshire had been

> ... open and covered with a beautiful coating of turf, profusely decorated with *Anemone pulsatilla* [pasque flower], *Astragalus hypoglottis* [purple milk-vetch, now *A. danicus*], and other interesting plants. It is now converted into arable land, and its peculiar plants mostly confined to small waste plots ... Even the tumuli, entrenchments, and other interesting works of the ancient inhabitants have seldom escaped the rapacity of the modern agriculturalist, who too frequently looks upon the native plants of the country as weeds, and its antiquities as deformities.

In the East Anglian fens a rich wetland environment, with its marshes and pools and bogs, teeming with wildlife, was almost entirely destroyed by enclosure and drainage and only in a few isolated places, most notably at the National Trust's Wicken Fen in Cambridgeshire, can we still experience the watery richness of the old fen commons, and look down, from the margins of the reserve, onto the bleak and

Enclosure and agricultural improvement were frequently to the detriment of traditional heath and downland flora. The pasque flower was a notable victim, and is now a scarce and local species in Britain.

desiccated surface of the surrounding arable land. Even in areas unaffected by enclosure and reclamation, the gradual simplification of the landscape – the constant tidying up of inconvenient field corners and of outgrown hedges – took its toll. In the 1870s the Norfolk rector and writer Augustus Jessop bemoaned the recent changes in the countryside:

> The small fields that used to be so picturesque and wasteful – where one could botanize with so much interest and pick up all sorts of odd pieces of information – have gone or are rapidly going; the tall hedges, the high banks, the scrub or the bottoms where a fox or a weasel might hope to find a night's lodging, all these things have vanished …

And there were social costs, too. In most parts of Britain farms had been getting larger, and fewer, for centuries, but parliamentary enclosure often hastened the process. For small owner-occupiers the allotment received in lieu of field land and common rights was seldom equal in value to what had formerly been enjoyed, especially once the legal costs, and the costs of fencing the allotment, had been taken into account. Many small proprietors sold out immediately to their more prosperous neighbours. Cottagers and labourers, however, fared worse. Many had habitually used the commons without having any strict legal right to do so. They received no compensation at all, other than perhaps the opportunity to benefit from some small 'fuel allotment' administered by a committee comprising the lord of the manor, the vicar and members of the local middle class. The effects of enclosure were most keenly felt where commons had been most extensive. But even in the champion Midlands, where common grazing had often been of limited extent, the poor lost out. In the Northamptonshire village of Raunds, for example, the 1797 enclosure was fiercely opposed by the cottagers and others on the grounds that they would be

> … deprived of an inestimable Privilege, which they now enjoy, of turning a certain Number of their Cows, Calves, and Sheep, on and over the said lands; a Privilege that enables them not only to maintain themselves and their families in the Depth of Winter … but in addition to this they can now supply Graziers with young or lean stock …

Where – as was often the case in the Midlands – enclosure was associated with the expansion of pasture, there were additional effects. Grazing required far less labour than corn growing. One Leicestershire contemporary noted the change:

> Before the fields were enclosed they were solely applied to the production of corn; the Poor had then plenty of employment in weeding, reaping, thrashing etc., and would also collect a great deal of cash by gleaning, but … the fields being now in pasturage, the farmers have little occasion for labourers and the Poor being thereby thrown out of employment, must of course be supported by the parish.

Hedges and walls

Hedgerow of an English Field by the landscape painter Berenger Benger, *c.*1905. As England became an increasingly urban nation in the course of the nineteenth century, the everyday features of the countryside were regarded with increasing affection and nostalgia.

Today, most people value hedges and walls as refuges for wildlife, or as essential features of unspoilt rural scenery. In the eighteenth and nineteenth centuries they could have other connotations. In his poem *The Mores* John Clare contrasted the freedom of movement enjoyed by the inhabitants of Helpston before enclosure, and the new landscape, in which:

'Fence now meets fence in owners little bounds
Of field and meadow large as garden grounds
In little parcels little minds to please
With men and flocks imprisoned ill at ease'

Soldiers clearing scrub, invading from the adjacent hedgerows, in 1917. The First World War provided a short interlude during the long agricultural recession of the late nineteenth and early twentieth centuries.

The Agricultural Recession

Throughout the middle years of the nineteenth century farmers and landowners invested heavily in new farm buildings, improved methods of land drainage, machinery, manufactured animal feed and artificial fertilisers. The Victorian age saw, in effect, the birth of modern, industrial farming. But from the late 1870s prices began to fall, and agriculture to slide into a long period of depression. The principal cause was the extension of the American railway network into the prairies of the Midwest, a prime wheat-growing region. European markets were flooded with cheap grain. Prices, rents and land values all fell steeply during the course of the 1880s and, after a brief respite at the end of that decade, a further intense depression occurred in the 1890s. This affected not only arable farmers but also livestock producers, as cheap meat and dairy produce were imported on refrigerated ships from Australia and the New World. During the First World War there was some revival of agricultural fortunes, but peace brought a renewed slump which continued, with only slight improvement, until the start of the Second World War in 1939.

Hedges and walls

The years of agricultural depression saw land coming out of use and the spread of nettles and other invasive plants. Whilst of interest to the botanical artist, they were no more than aggravating weeds to farmers.

The effects of this depression on the fabric of the countryside were considerable. Firstly, the area of arable contracted, reversing the trend of the previous century and a half. In 1869 6.2 million hectares were under cultivation in England and Wales, but by 1900 this had fallen to 4.9 million, and by 1939 to 3.6 million. Grass expanded at the expense of arable because livestock prices tended to hold up better than those for cereals, and because large urban conurbations continued to provide a good market for fresh milk. Secondly, there was a widespread retreat from the poorer soils, especially the heaths of the south and east and the high moors of the north and west. By the 1940s L. Dudley Stamp was able to describe 'fields formerly enclosed and cultivated which are being allowed to revert to vegetation resembling that of the unenclosed moorland by which they are frequently bordered'.

On the whole the depression years were beneficial for wildlife. The neat and manicured landscape of the 'high farming' years was replaced by something rougher and wilder. With farm profits low and labour costs relatively high, the slow but steady 'rationalisation' of boundaries which had continued in many arable districts through the late eighteenth and nineteenth centuries now came to an abrupt end. Moreover, hedges appear to have been cut and laid less frequently, especially in arable areas, and often grew tall and wide, providing better habitats for nesting birds. Less intensive management allowed many saplings to grow into trees, and the number of farmland trees (most of which grew in hedges) increased dramatically, from around 23 million in England in c.1870 to around 60 million in 1951.

But neglect, especially over a prolonged period, was not necessarily beneficial to hedges. Where they were not only allowed to grow tall, but were also subject to intense grazing pressure – especially by sheep, and particularly where they were composed largely of hawthorn – they could develop into disconnected lines of individual trees. Failure to maintain hedges in the traditional manner was further encouraged by the increasing use of barbed wire which, invented in America, was introduced into Britain in the 1880s. In livestock areas this was often employed to supplement deteriorating hedges or crumbling drystone walls, as well as to create entirely new fences. Other 'traditional' practices declined in this period. Pollarding now fell completely out of fashion and disappeared from most southern districts, although the practice continued sporadically in parts of the north and west. It had become increasingly uneconomic now that coal was cheap and easily available everywhere.

Recovery

The outbreak of the Second World War heralded a rapid change in agricultural fortunes, and major alterations to the fabric of the countryside. The sudden need to produce more food in the face of enemy blockade was followed by a further period of shortages and rationing which lasted well into the 1950s. Government money poured into the farming industry, and Britain's entry into the European Economic Community in 1973 brought a new range of subsidies. Farmers were now encouraged to produce more food, more intensively, which caused important changes in the practice and organisation of farming. The area under arable

The advent of mechanisation, and especially of large machines like combine harvesters, was an important factor in the large-scale hedge removal of the second half of the twentieth century.

A wheat 'prairie' stretches to the horizon in what was once the green Leicestershire countryside.

cultivation expanded once again, especially in eastern England – often onto light, infertile soils – and also in many parts of the Midlands, where grassland had become dominant during the eighteenth and nineteenth centuries. At the same time, the widespread adoption of tractors and combine harvesters, coupled with the availability of artificial fertilisers, ensured that many farmers no longer had the incentive, or the necessary resources, to invest in both livestock and cereal farming. Over considerable areas of England the numbers of livestock fell drastically, and even horses were, by the late 1950s, no more than a memory on most farms. On holdings which grew nothing but arable crops, walls and hedges were no longer required as stockproof barriers. More importantly, the new machinery worked most efficiently and economically in large fields. Hedges were increasingly seen as a nuisance, as a relic of old-fashioned and inefficient farming methods. Small fields and large hedges waste space: with fields averaging 2 hectares, even with hedges only 2 metres wide, some 2.6% of ground area is lost to crops; when field size rises to 40 hectares, this figure falls to only 0.6%. Hedges shelter vermin and shade crops, especially as previous decades of neglect had ensured that many had grown tall and wide. Buckthorn harbours crown rust, *Puccinia coronata,* spindle the broad bean aphid, *Aphias fabae.* Moreover, farms increased steadily in size, especially in arable districts, thereby reducing the number of perimeter fences required to define separate properties.

The removal of hedgerows was thus resumed, especially in the east of England, but now on a scale far in excess of anything seen before. It takes a group of labourers many days to grub out a hedge. It takes a bulldozer a matter of hours. Even where hedges were allowed to remain they frequently suffered neglect. They were redundant features of the landscape, maintained without enthusiasm. Many were badly damaged by stubble burning – a practice only made illegal in 1989.

It has been estimated that between 1946 and 1970 around 4,500 miles of hedge were removed each year in England and Wales, with the greatest losses occurring in the eastern counties and with less destruction in the west: in livestock farming areas hedges still had some purpose and were less of a hindrance to farming operations. Norfolk thus probably lost around a half of its hedges in this period, whereas Devon lost only around a tenth. But even within the east there was much variation. On the whole, large landed estates – where these had survived the recession years intact – tended to retain their hedgerows better than freehold farms, which were often the property of former estate tenants, men with less interest in country sports and who were keen to demonstrate their enthusiasm for modern farming. Large institutional landowners like pension funds, farming their land directly or by contract, probably had the worst record, removing anything that stood in the way of profit. And on the whole, areas of ancient countryside were more seriously affected by hedge removal than regions characterised by later, and especially parliamentary, enclosure where the large rectangular fields were more suited to modern farming.

Many hedges were badly damaged by stubble burning before the practice was made illegal in the late 1980s.

The rate of destruction also varied over time. In Norfolk for example, a particularly well-studied county, the scale was initially modest, with around 500 miles grubbed out each year between 1946 and 1954. It then accelerated, reaching c.2,400 miles per year by 1962, in part because hedge removal was subsidised after 1957 by the Ministry of Agriculture's Farm Improvement Scheme. The peak came during the following four years, when 3,500 miles were removed each year, the rate gradually falling thereafter to around 2,000 miles by 1970, the year in which subsidies for removal were withdrawn.

After c.1970 hedge removal seems to have continued at a lower rate, in part because on many farms in the east of England there were precious few hedges left to remove. But in part it was because opposition by conservationists and others was now reaching a crescendo. Whereas, in the immediate post-war years, the main concern of government and populace had been the production of large quantities of cheap food, by the 1970s food shortages were a thing of the past and the countryside was increasingly occupied by vocal members of the middle class, keen to maintain the kind of landscape which had attracted them to the countryside in the first place. Popular writers like Richard Mabey and Marion Shoard gave dire warnings of the effects of continued hedge removal. As Shoard prophesised in 1979:

> Hedgerow removal is in the end likely to be accompanied by the disappearance from many parts of our countryside even of common species, from wild roses to celandines and badgers to dormice.

By 1982, government policy had begun to change: under the Farm Capital Grant Scheme subsidies were offered for planting new hedges and for renovating outgrown, neglected ones. Nevertheless, even in the 1980s the destruction of hedges continued in eastern districts, often as small, old-fashioned farms came on the market and were absorbed into some larger, more commercially viable enterprise.

Meanwhile, in the Midlands and west of England the scale of removal, negligible in the immediate post-war period, speeded up in the 1970s and 1980s. As already noted, hedges were retained on a larger scale, and were better maintained, in stock-farming districts. But the frontier of arable cultivation expanded gradually westwards during these years. Even where stock farming remained important, hedges were often removed to make larger fields, or replaced with wire fences to reduce maintenance costs. The fashion for 'paddock grazing', involving monocultural grass swards and moveable fences, rendered both ancient hedges and ancient pastures redundant. Where hedges remained, their condition tended to deteriorate as knowledge of traditional management skills declined: often the lines of barbed wire or fences of large mesh sheep netting erected beside them were now the real stockproof barrier. Much the same was true of drystone walls in upland districts, and here too the problems of maintenance were compounded by

A hedge being grubbed out in Norfolk in the 1970s. The arable counties of eastern England, and East Anglia especially, bore the brunt of twentieth-century hedge removal.

the fact that the number of workers with knowledge of the skills required to keep them in good condition had steadily declined.

Changes in agriculture were not the only developments inimical to traditional field boundaries. Many were lost in the inexorable expansion of suburbs and industry; and many hedges were badly affected by Dutch elm disease. Contrary to what has sometimes been suggested, this was not an entirely new phenomenon (there had been earlier outbreaks, going back to prehistory, including one in 1927) but the strain of the disease which appeared in the late 1960s was particularly virulent. As a tree, the elm largely vanished, leaving many gaps in hedges and radically changing the appearance of those districts in which it had been the dominant hedgerow tree. Yet, as a shrub (and a hedge plant) the elm survives in vast numbers. Only when the plant acquires a substantial stem and a well-developed bark does it fall prey to the disease, although even then new growth can sucker from the root system. Even in relatively disease-free areas, the numbers of trees in hedgerows steadily declined. Some were deliberately removed, others fell victim to old age or storms: but either way they were seldom replaced in an age in which timber was seen as something to be grown apart, in special plantations, rather than being scattered across the countryside, where its presence might interfere with the efficient practice of agriculture.

During the 1990s the rate of deliberate destruction of traditional field boundaries slowed. But everywhere hedges and walls were, to varying extents, rendered redundant by changes in the practice and technology of farming, and their survival often rested on the whim of individual landowners and in particular their enthusiasm for hunting and shooting. Now we appear to be at the start of a new age, with hedgerows actually protected by legislation, and with grants for the planting of new hedges, and the maintenance of old, becoming increasingly available.

Hedge removal has not completely ceased. This hedge was destroyed without permission by the former National Trust tenant at Labour-in-Vain Farm in Dorset as recently as the late 1990s.

Chapter four

❖

Understanding hedges

UNDERSTANDING HEDGES

Hedges can display a bewildering degree of variation, and the more we come to know and notice them, the more this becomes apparent. Some of this variety is transient: the height and width of a hedge can be transformed dramatically in a day, with the help of modern machinery. But other aspects, such as the types of tree and shrub which form the main body of a hedge, change more gradually, and can give clues about its origins and history.

Hawthorn or may is a familiar hedgerow sight and at the heart of many country legends and superstitions. Despite the plant's long association with springtime renewal and fertility, it was traditionally considered bad luck to bring may blossom indoors.

The Composition of Hedges

A wide range of species can form the body of a hedge. Some, left unmanaged, will grow into sizeable trees, including ash, oak, the various kinds of elm, holly, sallow, whitebeam, goat willow, wild cherry, rowan, hornbeam, poplar (black, white or aspen), crab apple, beech, alder, hawthorn, field maple and sycamore. Others always remain as relatively low-growing shrubs, such as buckthorn, bullace, sloe, hazel, elder, dogwood, guelder rose, privet, wayfaring tree, and the various kinds of rose – dog rose, burnet rose, field rose and sweet briar. Most are natives, but some exotic introductions can be found. Sycamore is now common, especially in the west of England, while horse chestnut is sporadically encountered; lilac is a familiar feature of hedges on the sandy soils of coastal Suffolk, fuchsia and tamarisk in parts of the south-west.

As we shall see, a number of species have, at various times in the past, been used as hedging plants, but the most popular has always been hawthorn, also known as may or whitethorn. Indeed, its most common name – from the Old English *gehægan*, 'to enclose', the same word that has given us 'hedge' – attests its ancient importance in this role. Hawthorn grows quickly, flourishes on almost all soils, and is armed with particularly unpleasant thorns: it was made for hedging. Sloe or blackthorn was always the second most popular hedging plant. It has similar virtues but, being a strongly suckering species, it has the undesirable tendency of spreading into the adjacent field, gradually creating very wide and unmanageable hedges.

The mix of shrub species found today in particular hedges can display considerable variation. Many hedges are dominated by a single species, usually hawthorn, less commonly sloe, with other shrubs relatively limited in numbers and variety. At the other extreme we find hedges which are much more mixed and in which no species is dominant, other than for a few metres of length. Some of these variations reflect local environmental conditions. Most, however, are related to the origins of the hedge, to its antiquity, and to the ways in which it has been managed since it was planted.

Hedge Management

The most common method of maintaining a hedge has always been by laying or plashing, normally carried out in the harsh conditions of winter. Firstly, the hedge is hacked back with a billhook and the lateral suckers are removed, restoring it to its true line: dead material, thick old trunks and unwanted species – such as elder, which provides a poor barrier to stock – are removed. Next, the stems of the main shrubs are cut three-quarters of the way through, at an angle of 45–60°, around 5–10cm above the ground. The main stems, or 'pleachers', are then bent downwards at an angle of at least 60° so that each overlaps with its neighbour. As growth resumes in the spring a thick wall of vegetation is formed.

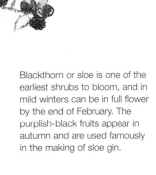

Blackthorn or sloe is one of the earliest shrubs to bloom, and in mild winters can be in full flower by the end of February. The purplish-black fruits appear in autumn and are used famously in the making of sloe gin.

Plashing takes a number of local and regional forms, but the main distinction today is between the Midland style and that of Wales and the South West. In the former, the hedge is drastically cut back before the remaining shoots and branches – the 'pleachers' – are laid. They are bent over and woven around poles of ash or hazel ('stabbers') which are spaced at intervals of about two-thirds of a metre. The bushy or 'brush' side of each shrub is laid so that it faces the field side of the hedge – that is, away from the associated ditch – so that the growth is protected from browsing. 'Hethers' or 'binders' – long, straight rods of elm or hazel, sometimes supplemented with saplings left for the purpose and bent at right angles – are then used to tie the stakes and to keep the 'pleachers' in place, forming a kind of continuous 'cable' along the top of the hedge. There are, or were, many local and regional variations in Midland practice. In some areas, notably Bedfordshire, it was once usual to lay each side of a hedge in turn, after a gap of several years, to ensure that it remained stock-proof.

A number of tools were employed in laying and coppicing hedges, including (below, from left to right) the maul, slasher and billhook. Both the slasher and billhook displayed a wide range of regional forms; the slasher was used to cut back surplus vegetation and the billhook to cut partway through the stems that were to be retained and laid. The maul, a kind of heavy wooden mallet, was used to knock in the poles or stabbers.

In Leicestershire the hawthorn hedges were laid in such a way that they grew very tall and thick – the so-called 'bullfinches', or bull fences, designed to contain the beef cattle which, after enclosure, became a particular speciality across much of the county.

In contrast, Welsh hedges, and those found across much of the south-west of England, tend to be laid lower and more densely than those in the Midlands, largely because they are designed primarily for sheep and not cattle: sheep are better than cattle at scrambling through gaps in the bottom of a hedge. Again there are many local and regional variations in practice, but in general the hedge is less drastically cut back before laying, and the 'brush' is often laid so that it projects on both sides of the hedge alternately – that is, it is 'double brushed'. 'Crooks', crook-shaped stakes, are often employed to hold down the pleachers. Brushwood may be added to the base of the hedge, in order to provide an additional barrier and protect the new growth; many hedges in the west, although they stand on prominent banks, have no lateral ditch.

Laying is an immemorial practice. Archaeological evidence suggests it was common in prehistoric times. It was certainly regarded as the usual method of management by early agricultural writers. In 1598 John Fitzherbert included a chapter on 'How to plashe or pletche a hedge' in his farming manual, the *Booke of Husbandrie.* This gives instructions on how to lay a newly planted hedge, at twelve years' growth; how to lay a well-established hedge; as well as the more severe treatment required to deal with outgrown hedges, comprising 'great stubbs or trees, and thinne in the bottome that Cattell may go under or betweene the trees'. In some periods, and in some places, hedges were managed by coppicing – pollarding at ground level. This is a practice usually associated with ancient woods, in which the majority of trees and shrubs were cut back to a stump or stool every ten to fifteen years, in order to provide a

A recently laid hedge in Gloucestershire, 1930s. Hedge-laying was a time-consuming and skilled job, and one that was easily dispensed with when agricultural recession hit.

regular crop of straight poles, suitable for firewood and other domestic uses. When applied to hedges, the method simply involved cutting all the shrubs to within a few inches of the ground at intervals of between ten and twenty years. Coppicing required less skill than laying and produced more useable firewood. But it demanded careful management, in that the new growth required protection from browsing livestock for two or three years. Where substantial ditches accompanied a hedge this was achieved by managing the adjoining land in such a way that animals were excluded from the unditched side for the necessary period or, alternatively, the hedge might be temporarily protected with hurdles or lines of staked brushwood.

The geographical distribution of laying and coppicing probably changed over the centuries and by the late eighteenth century was complex and without any very clear explanation. For obvious reasons coppicing was important in many primarily arable districts – much of East Anglia and parts of Bedfordshire, for example – but it was also widely practised in some livestock districts, such as Lancashire. The situation is further complicated by the fact that there were local traditions of hedge management which combined elements from both practices. In Middlesex, for example, it was customary to cut down the hedges every ten to twelve years 'to within a few inches of the bank'. A 'very thin hedge' was then formed from a few remaining stems, supplemented by stakes, and within two or three years the vegetation had recovered enough to provide a reasonably stock-proof barrier. In some places by the eighteenth century hedges were only laid or coppiced when they became gappy, and the hedge was otherwise maintained by regular trimming. But this was seldom traditional practice because it produced little firewood. In the words of one contemporary, hedges managed in this way 'cease to be the collieries of the country'.

This is an important reminder that although hedges were primarily intended to provide a stock-proof barrier and to define property, they supplied farmers with other benefits. They gave some shelter to livestock and crops, and – as we have seen – provided a variety of raw materials and products which might in turn affect the species planted, or allowed to establish themselves, within hedges.

We should also note that hedges are often accompanied by ditches, especially in districts with poorly draining soil, and that these were as important as the hedge itself. Together with the bank on which the hedge grew, the ditch helped provide a secure barrier against livestock but, more importantly, it helped drain the adjacent fields, and was connected to natural watercourses via a maze of similar drains. Ditches were cleared out every five years or so, a particularly unpleasant task carried out in the winter months. The excavated material was usually dumped on the hedge bank, which was thus augmented and strengthened. This explains in part the traditional legal position concerning the precise line of rural property boundaries, which are considered to run not along the line of the hedge, but on the side of the ditch. The ditch, in other words, is deemed to be parcel of the property on the other side of the associated hedge.

Oaks – shown here in a hedge in Warwickshire – are probably the most common hedgerow tree in England.

Hedgerow Trees

Trees – both timber and former pollards – are a feature of most, although by no means all, hedges. As a general rule, pollards are absent from those hedges established after c.1780, and timber of any kind is generally sparse in hedges planted after c.1820. As already noted, oak, ash and elm have – throughout the post-medieval period – been the most common hedgerow trees. But they are not, and apparently never were, equally common in all districts. Across much of the Midlands, especially in the grass 'shires' of Northamptonshire and Leicestershire, ash tends to be as common as oak, whereas in Norfolk and Suffolk oaks tend to dominate the skyline. In parts of Essex and Hertfordshire, especially on the poorer soils formed in the London Clay and associated Eocene deposits in the south of these counties, elm was the dominant hedgerow tree until the impact of Dutch elm disease in the late 1960s: the epidemic had a particularly severe effect on the landscape of these counties.

Outgrown beeches on a hedge bank at Chapel Steep on the Holnicote estate. Beech hedges, and beech windbreaks, are a characteristic feature of the Exmoor area.

Various other species were, and are, allowed to grow into hedgerow trees. Field maple can often be found, both as a standard and as a former pollard. Aspen, beech, willow, holly and hornbeam were locally popular, the last mainly in the South East, and especially in Hertfordshire and Essex. Its wood is extremely hard and was used for flails and mill machinery, although like the other species it was mainly pollarded for firewood. In the north and west of England sycamore, most probably introduced into this country in the sixteenth century, seems to have become established quite rapidly in hedges and is now common. Indeed, in parts of Cornwall it is the only tree to be found growing on the tall 'hedges' of stone and earth. On Exmoor, beech standards are a distinctive feature of the beech hedges established when the moor was enclosed and reclaimed in the nineteenth century. In parts of the Vale of Aylesbury in Buckinghamshire the black poplar is still the dominant tree in hedgerows near, or on, floodplains, mainly taking the form of ancient, gnarled pollards. Quite why this tree, common in medieval times but now rather rare in the rest of Britain, should remain important in this rather restricted area remains a mystery.

The age of hedgerow trees is often hard to gauge. As a very rough guide, free-standing timber trees (especially oaks) gain approximately two centimetres of circumference for each year of growth. Those growing in hedgerows, however, appear to girth at a slightly slower rate, largely because of competition with neighbouring shrubs (especially in the early years of growth). Either way, the majority of timber trees in British hedges seem to be less than 200 years old, probably because – before the second half of the nineteenth century – they were systematically exploited and felled as soon as mature, at around 75–100 years. Former pollards are usually much older, but their precise age is even harder to ascertain. Pollarded trees put on girth at a much slower rate than other trees, mainly because the periodic removal of the crown reduces the ability of the tree to grow, at least until a new crown of leaves becomes established. Some may be more than twice the age of free-standing trees of equivalent circumference. Almost all are more than 250, and many more than 500, years of age.

Where hedges were removed in the past, hedgerow trees were often allowed to remain, and such 'relict hedges' are an important, if subtle, feature of the landscape in a number of places. When ornamental parks were laid out around

Ancient oak pollards in Ickworth Park, Suffolk. Like many of the older trees in English parklands, these originally grew in a hedge.

the homes of the wealthy in the seventeenth, eighteenth and nineteenth centuries it was usually at the expense of enclosed countryside. Landowners grubbed out the hedges to create the required prospects of sweeping, open turf but left many of the trees standing, in order to provide their new landscapes with an instant sylvan appearance. Lines of trees – older than the park itself, and standing on slight banks marking the line of the former hedge – are a common sight in landscape parks, great and small. Particularly good examples can be seen in the fine parks, now owned by the National Trust, at Ickworth in Suffolk and Kedleston in Derbyshire. More surprisingly, lines of former hedgerow trees often survived where farmland was replaced by suburbia. Sometimes, as at Barn Hill in Wembley in the 1930s, the planning authorities actively safeguarded existing trees as the land was developed. More generally, developers retained trees simply because they enhanced the attractiveness and therefore the value of the properties on new estates. Old trees in suburban areas can usually be correlated with field boundaries (or even individual hedgerow trees) depicted on the First Edition Ordnance Survey maps from the late nineteenth century. They are of considerable ecological importance, providing in particular homes for the limited urban populations of barn owls.

Dating Hedges

The idea that the character and variety of plants contained in hedges can inform us about their age and origins has attracted considerable attention over the last three decades or so. In the 1960s the ecologist Max Hooper, working at what is now the Centre for Ecology and Hydrology, became intrigued by the extent to which the species composition of hedges displayed a wide range of variation. He concluded that age was the most important factor determining the range of shrubs found in a hedge, being more significant than aspects of soil or climate. After studying a sample of 230 hedges in Devon, the East Midlands and East Anglia, he developed an equation which could be used to estimate the approximate date of a hedge:

Age of hedge = (no. of shrub species in a 30-yard length x 99) - 6 years

or, in simplified form, 'one species = one century of age' (the equation excludes climbers like brambles and honeysuckle).The principal explanation given for this observed relationship was that hedges acquire new species over time, but fairly gradually, and at a roughly constant rate.

At around the same time Hooper's co-worker, E. Pollard, was developing a rather different approach. He noted that in some hedges, normally those containing a large number of shrub species, plants can be found which are closely associated with ancient woods and which only colonise new areas of woodland very slowly. These 'indicator species' include a number of perennial herbs growing at the base of the hedge, notably wood anemone, primrose, and greater burnet saxifrage. Dog's mercury and bluebell have also been considered as plants of this type, but in some districts these seem to colonise more rapidly and occur in many old hedges. A number of shrubs are also 'indicators', particularly small-leaved lime or pry and wild service tree. Neither species can propagate easily by seed under modern climatic conditions, only by suckering, and so they are largely restricted to areas of ancient woods. Woodland hawthorn, (*Crataegus laevigata*, distinct from the more common *Crataegus monogyna*), is also closely associated with old woodland, although it is more widely distributed in hedges. This is largely because it made a good hedging plant and was widely dug up and used for this purpose.

It was originally suggested that 'woodland relict' hedges were created by medieval assarting: ie lines of scrub were maintained while land was being cleared from the wilderness, layed and managed as a hedge. Some 'woodland relict' hedges may have come into existence in this way, but most seem to contain these distinctive plants simply because they were first planted next to areas of woodland. Some are found on major boundaries of parishes or ancient manors. These were perhaps originally laid out through what was only partially cleared woodland, in order to mark the bounds of the areas of waste exploited by separate owners or communities, presumably in Anglo-Saxon times. Woodland species would have colonised such boundaries with ease. Others seem to mark not the hedges created by assarting from extensive untamed wastes, but rather the former bounds of managed woods which survived well into the Middle Ages within cleared and settled landscapes, before being grubbed out.

'Hooper's Hypothesis' was enthusiastically embraced by local historians and others in the 1970s and 1980s, but was often applied rather uncritically. Hooper himself had been at pains to emphasise that the 'hypothesis' provided only a very approximate guide to the age of a hedge, but a number of criticisms of the method were soon made, most notably in an important article by C. W. Johnson, who pointed out that even when applied to Hooper's own set of data 90% accuracy of the age/species number correlation could only be achieved by allowing a margin of error of 200 years either way, rather than the 100 originally suggested. This rather lessened the usefulness of the theorem: a hedge with four shrub species might have been planted in the sixteenth century, but could just as easily have been established in the fourteenth or the eighteenth! Other writers drew attention to the statistical problems inherent in applying, as Hooper did, the technique of regression analysis to a set of data which, to use the jargon, has a bimodal distribution: ie in which there are many relatively young hedges (up to 300 years) and a number of very old ones (900–1100 years), but few planted in the intervening period.

In addition, a number of empirical studies, such as that carried out by Willmot in the parish of Church Broughton in Derbyshire, cast doubt on the extent to which species diversity was primarily related to age. Other environmental factors were, if anything, more important. In spite of such warnings 'hedge-dating surveys' are still carried out in many districts. But there is now little real doubt that, while in general old hedges have more species than recent ones, the correlation between age and species numbers is nothing like strong enough to be used as an independent, accurate dating method; and indeed, that even the more general relationship which exists between age and species content has some important exceptions.

Perhaps the most important reason for this is that hedges could be, and often were, planted with a range of species rather than with just one. This was particularly the case, as we have seen, in the period before the late eighteenth century. Pehr Kalm, who visited England in 1748, typically noted how in the Hertfordshire Chilterns hedges were commonly planted with a mixture of hawthorn and sloe, but that in addition the farmers 'set here and there, either at a certain distance or length from each other, or just as they please, small shoots of willows, beeches, ash, maple, lime, elm, and other leaf-trees'. In other words, to some extent old hedges are species-rich simply because they were originally planted with a greater range of shrubs and trees than those more recently established. Indeed, it is noteworthy that two centuries ago agricultural writers were already making the same broad distinction so apparent today: between the hedges of hawthorn, which were then in the process of being planted as open fields and commons were being enclosed; and older hedges, which contained a range of species. But even in the nineteenth century smaller landowners, especially those living in anciently enclosed districts, sometimes saved money by planting

View across the Olchon Valley in Herefordshire, showing mixed hedges dominated by hazel, maple and hawthorn. The hedge in the foreground has grown gappy through poor management and intensive grazing.

whatever they could find in local woods and hedges when enclosing a piece of land, or included in the hedge shrubs and trees which they would find useful. The hedges created at the enclosure of Neroche Forest in Somerset in 1833 – with an average of 6.3 species per 30-metre stretch – are one well-known instance, but there are innumerable local examples. One researcher, Philip Lazaretti, recorded a seven-species hedge created at the enclosure of Denton Common in Norfolk in 1815! On the other hand, we should also note that hedges were sometimes grubbed out, and replanted on the same line with one or two species, thus effectively resetting their chronological 'clock'. This phenomenon is well-attested in the eighteenth and nineteenth centuries, but has been noted in medieval Yorkshire and was probably locally widespread at various times in the past, especially when periods of agricultural boom followed phases of recession and neglect.

In addition to all this, as Hooper himself warned, the availability of seed types can affect the rate at which new species become established in a hedge. Imagine a hedge planted in the middle of Salisbury Plain, an area largely devoid of trees and hedges since late prehistoric times, in 1300; and one planted at the same time in the Weald of Kent, an area full of woods and with numerous existing hedges. It is inconceivable that both should have acquired the same number of new species seven hundred years later. Such variations can work at a much more local level. Hedges near woods are frequently more species-rich than those only slightly further away, and may contain unusual species locally found only within the wood in question. And studies in several areas have suggested that roadside hedges often contain more species than those around fields, and this also might be a consequence of increased seed-supply, associated with the movement of livestock and traffic.

More generally, it is apparent that species can be lost from hedges over time, as well as gained. Elder is common in recent hedges but rare in old ones: it colonises rapidly but in a well-managed, densely-growing hedge is eventually forced out by other plants (returning only when gaps appear). Certain kinds of elm sucker along a hedge, displacing existing species, and in extreme cases can convert a medieval, multi-species hedge into one consisting only of elms. And in a host of ways the development of the hedge flora over time differs from the normal pattern of biological succession seen, for example, when abandoned open land gradually becomes wooded. This process involves the replacement of short-lived but rapid-growing species with herbaceous perennials, which are themselves supplanted by slow-growing, long-lived shrubs and trees. But hedges begin life in the latter stage and are sometimes planted with species that would not naturally co-exist. They are then managed to prevent their full growth into (in many cases) tall trees, while some species – poor hedging plants or those deemed a threat to crops – are systematically weeded out, such as barberry, dug out since at least the eighteenth century because it is a host for the fungus that produces black stem rust in wheat.

But perhaps most importantly of all, Hooper's method pays insufficient attention to the fact that variations in soil and climate can, and do, affect the number of plants that grow in hedges. Leached and acid soils, for example, can support a smaller range of species than fertile and alkaline clays. One otherwise enthusiastic hedge-dater, working in Essex, was thus able to declare that while Hooper's method worked well on the boulder clay and chalk soils of the county, it 'should be treated with reservation on certain soils; the acid Bagshot and Claygate beds, and areas of glacial outwash, encourage fewer species'. In the north of England and across much of Wales even very old hedges are often relatively species-poor, simply because climatic conditions mean that many common hedge plants are here at the edge of, or even beyond, their natural range. In Ireland, many hedge plants common on the British mainland are rare or absent altogether, including dogwood and field maple.

There are many other objections. Hedges are in a complex, dynamic relationship with both the environment and human society. They are cultural artefacts as much as natural features, and the history of an individual hedge is thus closely linked to the economic history of the farm on which it grows, as well as that of the wider locality. If hedges are poorly managed and allowed to grow wide and tall – as often seems to have happened during periods of agricultural recession – certain plants, including dogwood, can be lost from them. Pronounced shading, caused for example by the temporary development of scrub in an adjacent field, can have more drastic effects, converting a multispecies hedge into a line of pure hazel or hornbeam. With subsequent economic recovery the hedge might be tamed once more, and adjacent scrub removed, leaving no trace of the processes that had severely adjusted the hedge's composition.

Given all this, it is surprising that the hedge-dating equation was accepted by so many people for so long. Let me repeat: there is absolutely no doubt that hedges cannot be dated by simply counting the numbers of species they contain. Yet we must be careful not to throw the baby out with the bathwater. There is much that both the local historian and landscape historian can learn from the careful examination of hedges. We can begin with noting that certain species (hawthorn, sloe, rose, elder, ash) are relatively rapid colonisers; that others (hazel, dogwood, holly) are slow colonisers; while some (wild service, small-leaved lime, most types of elm) are very poor colonisers, propagating only or largely through suckering. Now it is true that these general, broad rules are modified by the fact that rates of colonisation are affected by the character of the local seed supply. Thus field maple is generally found only in the very oldest hedges in the Midlands, where the medieval landscape was largely hedgeless and treeless, yet it is moderately common even in parliamentary enclosure hedges in parts of East Anglia, where it was abundant in woods and hedges throughout the medieval period. Moreover, as we have seen, variations in initial planting, and in later management, undoubtedly affect species content. Yet for any district or region, peculiarities of environmental and agricultural history – of local traditions of planting and maintenance, of cycles of management and neglect – generally ensure that the shrub content of hedges is far from random, and can, at least in broad terms, be related to origins and age. In other words, categories of hedgerow can often be identified by the careful linkage of cartographic and documentary sources, and field observation; once identified, these can be used to throw light on the history of the local landscape in circumstances where these forms of 'hard' evidence are limited, or absent.

The claylands of south and central Norfolk, for example, have a complex range of boundaries. In medieval times the countryside comprised a mosaic of open fields, enclosed land and extensive open commons. The last generally occupied areas that were particularly poorly drained and became important foci for settlement from later Saxon times onwards. The open fields disappeared gradually from the late Middle Ages, enclosed piecemeal and often laid to pasture: few survived by the end of the seventeenth century. Some encroachments on the commons occurred over this same period, but they remained substantially intact until removed by parliamentary enclosure, mainly during the Napoleonic Wars. At the same time, there were numerous changes to the pattern of boundaries within enclosed ground, associated with a return to arable farming from the 1780s.

The hedges, not surprisingly, display considerable variation in species content, but most fall into one of six categories. Firstly, there are those created when the larger commons were enclosed, in the decades either side of 1800. Most are still composed mainly of hawthorn, but they also contain some elder and considerable

A medieval hedge near Littlebury Green in north-west Essex, featuring hazel, hawthorn, sloe, maple, dog rose, dogwood and wayfaring tree

quantities of sloe, rose and ash: this last, in a few hedges, has even become the dominant plant. Elm can also sometimes be found – probably suckering from the stumps of long-lost timber trees. Most hedges have four or five species per 30 metres suggesting, according to a literal interpretation of Hooper's 'rule', that they were planted in Tudor times!

Secondly, there are hedges of the same general date – later eighteenth or early nineteenth century – which were planted in slightly different circumstances: when the smaller commons in the area were enclosed; towards the margins of the larger commons; or when new boundaries were created within the surrounding, long-enclosed landscape. These hedges are broadly similar in composition to those just described, but also contain small quantities of other plants, including allegedly slow colonisers – particularly field maple but occasionally, where hedges abut ditches regularly flowing with water, even dogwood. Most have between five and six species in every 30-metre stretch: evidently, the availability of seeds from old hedges in neighbouring areas has here caused a major deviation from Hooper's 'rule'.

Even in areas of 'ancient countryside' some comparatively recent species-poor hedges can be found. This example near Lavenham in Suffolk is mainly composed of hawthorn.

Thirdly, there is a group of hedges with between five and eight species per 30 metres, but which are still clearly dominated – if only just – by that with which they were first planted: hawthorn or sloe. Hazel, maple, holly and elm are well-represented in these hedges which, to judge from cartographic evidence, were planted as open fields were enclosed, or intakes made from the commons, in the period between the *c*.1300 and *c*.1700. Some evidence suggests that this group can be broken down further into those of medieval and post-medieval date, but only within very broad limits of confidence.

A fourth group of hedges appears to have early medieval origins. These are characteristic of ancient common edges, parish boundaries and some roads, but also occur more widely in parishes in which documentary evidence suggests that a rather high proportion of land lay in enclosed fields from an early date. These are more mixed in composition than those so far discussed: hawthorn and sloe are less dominant, and there are often great masses of maple, hazel and (in particular) dogwood. In some places these plants can almost monopolise the hedge for as much as ten or fifteen metres: in consequence, the number of species per 30 metres is, somewhat paradoxically, often dramatically reduced. Indeed, the actual number of species in such hedges is sometimes less than that in boundaries established in the early nineteenth century.

The fifth category is by far the smallest: a tiny group of hedges similar to those just described but which contain, in addition, such 'woodland relict' species as small-leaved lime and wild service tree. These are found, in particular, on ancient hundred boundaries, and in places where areas of woodland were grubbed out and converted to fields in medieval or post-medieval times.

Lastly, we should take note of a type of hedge likely to cause much confusion in the near future. Recent hedges, established for nature conservation purposes, are usually planted with a wide range of species, some of which, such as guelder rose, are relatively rare in 'genuine' old hedges in the locality. Most still boast plastic guards and/or the remains of a mulch mat, but in a few years they will be hard to distinguish from medieval hedges!

A rather different range of categories was identified by Trevor Hussey in his classic studies of Emmington in Oxfordshire and Napshill in Buckinghamshire. The former village lies on fertile, loamy soils in the Thames valley. The parish was largely open-field land in the Middle Ages, but by late medieval times some enclosure had taken place: 'The species profiles show these older hedges to be distinctive, being rich in hazel, ash and dogwood, but poor in elm. They have an average of 6.2 species'. In 1697 the remaining open fields were enclosed by agreement. The new hedges, shown on a map of that date, had rather different profiles, with less hazel and ash but with much dogwood, maple and crab apple, and with an average of 4.4 species. Lastly, a number of hedges were added in the nineteenth century. These are characterised by a dominance of hawthorn, with some sloe, dog rose and ash and small quantities of elder and maple. In Napshill, high in the wooded Chilterns, there was less documentary or cartographic evidence but the hedges again fell into three broad categories. Firstly, there was a group of 'mixed species hedges with one or more woodland herbs enclosing large areas with rather sinuous outlines. They have an average of 6.5 species and their profile is distinguished by having more hazel than hawthorn, very little elder, an abundance of field maple and dogwood and some oak and beech'. A second group comprised mixed hedges with an average of 5.3 species, with less hazel than hawthorn, a high frequency of elder and holly, but little field maple, beech, oak or dogwood. Lastly, a small group of nineteenth-century hedges contained hawthorn, elder, ash and sloe, but very little else.

For any district or region such studies can throw important light on the history of the landscape, helping to distinguish (for example) the original, early medieval boundaries of Norfolk clayland commons from those created by later intakes. But such evidence should always be used with caution: exceptions and anomalies, sometimes quite inexplicable, always remain. In addition, researchers should always combine consideration of the hedge's botanical content with other aspects of the boundary's form and layout. Thus in many areas massive hedgebanks tend to suggest an early – often medieval – boundary, while eighteenth and nineteenth-century hedges, especially on lighter land, often have only slight banks. Ruler-straight boundaries are normally post-1700; gently curving or slightly sinuous ones usually derive from piecemeal enclosure in the fifteenth, sixteenth or seventeenth centuries; while hedges which follow an irregular line are probably of medieval origin. Hedges which run continuously for some distance, with other boundaries butting up against them, are often the earliest features in the landscape.

Locally Distinctive Hedgerows

While the age of a hedge is thus a major factor determining the kinds, and numbers, of shrubs it contains, other influences are also important. The character of the soil, as already noted, is of particular significance. Alder and willow are thus often found in hedges growing on waterlogged soils but are much less frequent in well-drained locations. In north Essex and east Hertfordshire buckthorn is only sporadically found in hedges growing on the boulder clay soils; in the major valleys, however, where the underlying chalk is more exposed, it is frequently encountered in older hedges.

Perhaps the most important aspect of all this is the broad distinction between hedges growing on lime-rich soils, especially damp calcareous clays, and those found on more acid, base-poor soils. The former tend to be more species-rich than the latter, even when of similar age, and this can produce profound regional patterns of variation. Thus in the area of 'ancient countryside' lying to the north and east of London there is a clear distinction between the hedges found on the chalky boulder clay and those on the London clay, sands and gravels. On the boulder clays, most hedges are mixed and species-rich, normally with between five and ten species in a typical 30-metre length. On the more acid soils the hedges are of comparable antiquity, and also mixed, but they contain noticeably fewer species, generally six or less in a 30-metre length. In part this is because large sections are occupied by

Hedgerow elms at Whaddon in Cambridgeshire. Although disease-free mature elms are rare, youngsters such as these are still common.

sloe or elm, which successfully sucker and out-compete other shrubs on these poor soils. This tendency increases markedly towards the coast and the Thames estuary, and across much of the south Essex plain and on the Dengie peninsula many hedges are now composed entirely of elm. The fact that the fields they define form extensive semi-regular grids, probably first laid out in late prehistory, is powerful demonstration of the fact that the number of species growing in a hedge can indeed provide a very poor indication of its antiquity.

Elm hedges occur in other coastal districts, most notably in northern Kent and parts of coastal Suffolk, but they are not confined to such locations. There are many in central Worcestershire and the Vale of Evesham, for example. As already noted, elm was used as fodder and this may have encouraged its use in hedges, although the suckering ability of most varieties and their ability to thrive in difficult locations probably explains most hedges of this kind. Elm hedges are one of several distinctive hedgerow types, some of which are widely but sporadically distributed, others restricted to particular localities. Hedges composed largely of holly – or with significant lengths of pure holly – can be found in a number of districts, including parts of south-east Suffolk (especially the Shotley Peninsula), in the area around Woburn and Ampthill in Bedfordshire, and in pockets on the Chiltern dipslope in Hertfordshire, especially around Sarratt and Chipperfield to the west of Watford, where they cluster noticeably in the vicinity of settlements. But they are principally a feature of the counties bordering the Pennines, from

Holly standards in a Herefordshire hedgerow. Only female trees bear berries, and flowering and fruiting can take place at any time of year. In some areas a holly tree was traditionally used instead of a fir or spruce at Christmas.

Nottinghamshire through to Staffordshire (especially around Glossop, Leek and Uttoxeter) and northwards through Cheshire to Lancashire; they are also a feature of some of the main Pennine Vales, including Nidderdale. The tall, wide, holly-dominated hedges which are a notable feature of Herefordshire and Warwickshire (the holly here generally well-mixed with a range of other species, including hawthorn, hazel, field maple, elm, ash and oak) are perhaps best considered as a south-western continuation of this northern holly province. In most cases, holly as a dominant species occurs in 'pure' blocks, of five to ten metres in length, within hedges that display varying degrees of species-richness. Less commonly, single-species holly hedges can occur, most notably in Bedfordshire but occasionally in Suffolk and Warwickshire.

In Bedfordshire, and perhaps in Suffolk, the abundance of holly may simply reflect the plant's ability to thrive on poor sandy soils. In both cases the hedges are restricted to sandy areas, on the Greensand deposits and the former heaths of the Sandlings respectively (holly, it should be noted, is a prominent feature of the only surviving ancient wood-pasture on the Suffolk Sandlings, Staverton Thicks in Wantisden). But elsewhere holly was perhaps favoured as a hedging plant for other reasons. As already noted, the plant was widely used as winter fodder, especially for sheep, and it is noteworthy that it is in these same districts – and especially in the triangle formed by Derby, Leeds and Manchester – that there is a high incidence of place-names featuring the elements 'hollin' or 'holly', attesting the antiquity of the practice. More importantly, holly makes a good stock-proof barrier which is very tolerant of grazing: new shoots appear from cut stems both above, and at, ground level.

Elm and holly hedges are found, for the most part, in areas of relatively early enclosure. But areas of late enclosure also have their distinctive hedges. Perhaps the most striking are those of Scots pine which surround the fields in parts of the East Anglian Breckland. Similar pine rows can be found more sporadically in other sandy regions, notably the Suffolk Sandlings. Only a very small number, mainly on the Elveden estate in Breckland, are still maintained as true hedges. All were established following the enclosure of these sandy lands at the start of the nineteenth century and the majority seem to have been planted by the owners of the large estates which dominated these areas. Locally known as 'deal rows' ('deal' being the old East Anglian term for a pine tree), the trees were probably selected for their ability to thrive on the poor sandy soils, and to form a dense shelter which provided both cover for crops and game and some protection against the erosion to which these light soils are prone. It is noteworthy, however, that they are not entirely restricted to the very worst of the Breckland soils, and can be found on the more calcareous and less unstable soils which also occur in the region. Presumably their planting became something of a fashion in the district in the early and middle decades of the nineteenth century. In fact, it is by no means certain that all were ever managed as true hedges. Some of the rows are composed of fairly straight, upright trees with little if any evidence of cutting

Hedges and walls

Typical pine row on the Cockley Cley estate in Norfolk's Breckland. The twisted shape shows that the trees were once laid as a hedge, their size suggesting that this form of management ceased within living memory.

The Duke of Argyll's tea plant originates in Asia, but was introduced to Britain in the eighteenth century. Despite its name, most parts of the plant are toxic.

and laying. Even some of those which do show clear signs of such management had already grown into trees by the time the First Edition Ordnance Survey 6-inch map was prepared in the 1880s.

Landowners and farmers enclosing land in the eighteenth and nineteenth centuries made use of other non-indigenous species, creating hedges which are now curiosities of the local landscape. In parts of west Norfolk, and on the Suffolk Sandlings, the Duke of Argyll's tea plant, *Lycium barbarum* was sometimes used as a hedging plant; along the south coast westwards from Hampshire to Cornwall, tamarisk was widely employed; fuchsia is found sporadically as a farm hedge across much of western, and especially south-western, Britain; while in parts of Cornwall and on the Scilly Isles, evergreen spindle and *Pittosporum* were planted in the nineteenth century around the small fields used to grow early daffodils. In and around the New Forest hedges of gorse can be found, some perhaps created by leaving a line of the plants when the surrounding heathy ground was cleared, others perhaps the result of colonisation along a fence line. It is said locally that families with smallholdings bounded them with gorse hedges and then pruned only the inside of the hedge regularly, allowing the outside to grow, thereby gradually extending the acreage of their plot!

Other locally distinctive hedges include those planted during the eighteenth and nineteenth centuries with laburnum, and a range of useful plants like damson, gooseberry and spindle, by squatters and free miners around encroachments from the waste in the old mining areas of south and west Shropshire. They are especially common in the vicinity of the villages of Shelve, Pennerley, and Stipperstones. These hedges break Hooper's rule with gusto: some 30-metre stretches contain more than twenty species. Damson – eaten as a fruit and once used to produce dye – makes for distinctive hedges elsewhere. Damson-dominated hedges are found widely across the West Midlands, in Herefordshire, Worcestershire, Shropshire, Staffordshire, and also in parts of Nottinghamshire. Again, most of these seem to be relatively recent additions to the landscape, of eighteenth or nineteenth-century date. More recent still are the hedges composed of wild plum (or bullace) and sloe, found in a small area of north Hertfordshire and associated with parishes which were never officially enclosed – the lines, although managed as hedges, seem to have developed spontaneously on old lynchet banks within the open fields.

A Herefordshire hedge composed largely of damson. In Kent damson trees were formerly planted as windbreaks for orchards, whilst in parts of the North and Midlands the fruits were used to produce dye for the textile industry.

Hooper's 'rule' may not have stood the test of time but his pioneering work has encouraged many botanists and historians to look at hedges with new eyes. We may not be able to date hedges by counting the number of species they contain, but botanical examination can, nevertheless, tell us much about the human, as well as the natural, history of an area.

FACING PAGE Hedgerows are a rich source of food for many types of wildlife. In spring magpies are a common sight as they hunt for eggs and young birds.

The dormouse's preferred habitat is hazel coppice, but the species often forages in hedgerows and there is evidence to suggest that hedges can form 'dormouse corridors', connecting otherwise isolated stands of hazel.

Hedgerow Wildlife

The mass of a hedge is composed of the kinds of shrub and tree species already mentioned. But its ecological value is dependent on more than just these core structural components. The complexity of a hedge, and in turn its potential to harbour a variety of wildlife, is enhanced considerably by the presence of other plant species, in particular climbers. The most common of these – and certainly the most noticed by visitors to the countryside – are brambles, loaded with fruit in early autumn, and perhaps cleavers (also known as goosegrass or sticky willie!).

Plants that climb rely on the stems of the bulkier hedge shrubs for support, climbing over and in some cases smothering them, especially where hedges have been poorly maintained: brambles in particular often occupy significant lengths of hedge where gaps have appeared following the loss of elms through Dutch elm disease. Some of these plants climb by twisting the growing tip of their stem to the right (hedge bindweed) or left (honeysuckle, hops and black bryony). Others climb by attaching themselves to the stems of plants through curling tendrils located at the base of the leaf stalks, like white bryony (black and white bryony are quite unrelated plants, the latter being a somewhat unlikely member of the gourd family, related to the cucumber). Old man's beard or traveller's joy, a member of the clematis family, has toothed leaves on long stalks which twist around and grip adjacent stems, while ivy clings to its support by using a number of tiny outgrowths which sprout from the shaded sides of its branches. Others still are scramblers, like woody nightshade: they climb by simply threading pliant stalks in and out of other plants. Collectively climbers provide a rich habitat web within the hedge proper, on which many other species of wildlife depend.

Climbers make an important contribution not only to the ecology of hedges but also to their appearance, and because certain species are more prominent on particular soils they can add much to the distinctive character of local hedges. On the rolling chalky hills of north Hertfordshire and Essex, for example, traveller's joy is a striking feature of the hedgerows, especially the larger and older ones. As early as 1548 William Turner commented how 'Wild clematis ... groweth plentifully between Ware and Barckway in the hedges, which in summer are in many places al whyte wyth the downe of thys vine'.

Most hedges, especially those in areas of heavy soil, have an accompanying ditch. This usually dries out in summer, and is characterised by such species as comfrey, hemp agrimony and meadowsweet. On the lower and damper parts of the banks, lords-and-ladies, campion (*Silene* spp.) and creeping buttercup are common,

Hedges and walls

whilst higher up, on drier soil, the flora is more akin to that of hedge bottoms, with yarrow, betony and hedge woundwort usually prominent.

The character of hedges changes dramatically with the seasons. In spring, their bases are alive with colour, as violets, primroses, celandines, lady's smock and jack-by-the hedge come into flower. In older examples wood anemone and primrose appear, while in some districts dog's mercury provides a brief flash of vivid green before being engulfed in grass and taller herbs. May is dominated more by whites and reds; by alexanders, cow parsley, red campion, greater stitchwort and lords and ladies, as well as the white snowy shock of the hawthorn blossom itself. In June a wide range of clovers and vetches bloom around the base of the hedges, while elder, dog and burnet roses and (more rarely) honeysuckle and guelder rose appear higher up. By late summer, hedge woundwort, foxglove, bramble, hogweed, yarrow, herb robert and many others are in flower. Month by month, hedgerows provide a constantly changing parade of colour.

Several of our wildflowers have names which proclaim their close association with hedges, such as hedge bedstraw, hedge parsley and hedge mustard. Rather more interesting are the local or archaic names for common plants: 'hedge bells' for bindweed in the Isle of Wight, 'hedge feathers' for old man's beard in Yorkshire, 'hedge grape' for bryony in Worcestershire, 'hedge-maids' for ground ivy in East Anglia. According to R. C. A. Prior, writing in 1879, the name of the common plant jack-by-the-hedge (also known as hedge garlic or garlic mustard) comes from 'Jack or jakes, latrina, alluding to its offensive smell'! Alas for so picturesque a tale: the plant only emits its (not particularly unpleasant) garlic-like smell when crushed, and 'jack' is a frequent term of familiarity applied to many plants. The greater bindweed was thus known as 'jack-run-in-the-hedge' in parts of western England.

Many verge species – such as dead nettle, or cow parsley – can be found in areas of waste ground or uncultivated land, but this does not lessen their importance, either visually or ecologically, in hedgerows. The larvae of the orange-tip butterfly, for example, feed on cuckoo flower, a plant which, because of the widespread destruction of its other main habitat – damp meadows – is in many areas now largely restricted to hedgerows.

Hedges support a bewildering range of invertebrates, which live not only on and in the plants and shrubs but also amongst the litter at the hedge base, and in the bank and associated ditch. For example, recent research by John Dover of Staffordshire University has shown that hedges are far more valuable for butterflies than was previously understood, with 39 species (64% of those on the British list) associated with hedgerow habitats. Twenty-six species are considered likely to breed in hedges, and these include the commoner types such as brimstone, orange-tip, green-veined white, comma, small tortoiseshell, peacock, holly blue, and the various kinds of 'browns' (the name of one of which, the hedge brown or gatekeeper, attests to its close association with this habitat). Other species, such as fritillaries, have breeding requirements that cannot be provided by hedgerows, but they will still use hedges for nectar, shelter and, occasionally, as transport corridors between patches of breeding habitat.

Some butterflies, such as the rare and local brown hairstreak, are hedgerow specialists: the hairstreak spends almost its entire adult life high up in the tops of hedgerow shrubs and trees, occasionally descending to feed on bramble blossom. Indeed, because the population is so widely dispersed – only one or two adults may emerge for each mile of hedge – the species makes use of certain 'master trees', where males and females congregate for breeding.

Essentially a meadow species, cow parsley is often found in great drifts along hedgerows.

The main value of hedges to butterflies may rest in their provision of 'surrogate' habitats for certain woodland and grassland species. Field corners, particularly those with a southerly aspect, offer useful sanctuary, and the physical configuration of hedges – with a lee side often sheltered from wind and rain – may assist butterflies to continue flying, feeding and mating during inclement weather.

Less obvious than butterflies are the invertebrates that dwell in and below the hedge. Some, such as the woodlouse, live among the decaying vegetation at the bottom of the hedge. Others feed mostly on the leaves of the main hedgerow shrubs, although these have often developed protective mechanisms which make them less palatable: the thin hairs on the leaves of the hazel, for example. Hawthorn, sloe and crab apple are especially important as insect food, but oaks support the greatest diversity of life, with as many as 300 species feeding on their leaves. Many insects, however, feed not on the vegetation of the hedge but on other insects: examples include lacewings, earwigs and ladybirds. Many of Britain's 40-plus species of ladybird are to be found in hedgerow habitats.

The brown hairstreak, shown here on fleabane, is perhaps more particularly associated with hedgerows than any other British butterfly (with the possible exception of the hedge brown).

Although commonly known as the hedge sparrow, the dunnock is not a member of the sparrow family at all. A shy bird, it becomes more obvious in spring when it can often be seen singing from a prominent perch.

One of the most interesting hedgerow
invertebrates is the glow-worm, actually
a type of beetle, which although rather
reduced in numbers over the last fifty years
or so, can still be found in suitable hedge verges
and overgrown banks in many parts of Britain.
Glow-worm larvae feed on small snails and slugs,
which restricts the distribution of the species to areas
of chalk and limestone – the snails need the calcium
for shell-building. Another classic hedgerow beetle is
the hawthorn shieldbug; in spring, the overwintered
adults eat the young hawthorn leaves, switching
their attention later in the year to the fruit.

Numerous species of bird feed on these insects, as
well as on the nuts, fruit and seeds produced by
hedgerow plants. Several of the most common British
birds flourish best where hedges survive in some
numbers. Magpie, wren, robin, chaffinch, blackbird, song
thrush and hedge sparrow are all closely associated
with hedges, as are less common but still familiar
species such as the yellowhammer and
bullfinch, and summer migrants like the
whitethroat. Birds more typically
associated with woodland – such as
certain members of the tit family – will also

Mixed thrush flocks, typically of
fieldfare and redwing, regularly
raid hedges for berries. Hawthorn
is a particular favourite.

Another common hedgerow
species, the comma butterfly
undergoes mysterious population
fluctuations. It is currently on the
increase, and extending its range
to the north and east.

A real hedgerow specialist, the tree sparrow has the unenviable distinction of being one of the greatest farmland bird 'casualties' of recent decades. Numbers collapsed, but are now showing signs of a limited recovery in some areas.

use hedgerows for feeding, roosting and breeding, particularly if the hedges contain well-grown shrubs and trees. The presence of so many smaller birds in turn draws the attention of predators such as the sparrowhawk, essentially a woodland bird, but one which can often be seen hunting along hedgerows. Large, overgrown hedges were formerly much favoured by the red-backed shrike, once a characteristic species in England and many parts of Wales, but now sadly extinct as a regular breeding bird. However, its demise may be due less to hedge removal than to other factors related to climate and conditions in its wintering quarters in Africa.

Species-rich hedges may provide a wider variety of vegetable food, but those that are species-poor often yield greater quantity. Elder, hawthorn and rose, for example, are often the dominant species in such hedges, and all produce large numbers of berries and haws, making them very attractive to birds in winter when other types of food can become scarce. Thrush species such as the redwing and fieldfare, which descend on Britain in large numbers in autumn and winter, use hedgerows extensively for feeding and roosting, and the lucky observer may even come across the exotic waxwing, an irregular winter visitor with a fondness for berries and rose hips. For birds especially, the way in which hedges are managed seems to have a vital effect on numbers: low-clipped, regularly-cut hedges support fewer birds than tall, spreading ones, cut back at longer intervals.

Yet although large, 'woody' hedges are preferred by most types of typical farmland bird, recent research indicates that there are certain species which do not conform to this pattern and which may do less well in areas where such hedges are dominant. In particular, red-legged partridge, skylark and meadow pipit – all open-country species – often show their greatest abundance where there is a preponderance of closely trimmed, treeless hedgerows dominated by a single species such as hawthorn.

It is not just the structure and composition of the hedges *per se* that are important for birdlife. The shelter provided by verges can be critical for birds such as the grey partridge, a nationally declining species which prefers to breed in the tangled vegetation on the edges of fields, close to the security of overhanging hedge cover. Excessively manicured hedges, combined with intensive field usage and the use of pesticides, have not helped the grey partridge.

Lastly, we should remember the large numbers of mammals which find shelter and food in hedges. Of the 28 species of lowland mammal in Britain, 21 breed in hedges, 14 commonly. The close association of one – the hedgehog – is proclaimed in its name, but the wood mouse, field vole, pygmy shrew and common shrew are all frequent hedge dwellers, and the bank vole, as its name implies, is also particularly closely associated with this habitat. Voles and mice normally run through the vegetation in shallow tunnels, but they can climb to reach fruit, especially rose hips: mice tend to remove the flesh and eat the seeds, voles to eat the flesh alone. Rabbits frequently make their homes in hedgebanks, and hedges – particularly those featuring hazel – continue to provide valuable habitat for the dormouse in those parts of England and Wales where this enchanting species still hangs on.

Of the larger mammals, badgers prefer to make their setts in woods, but they will also occupy large old hedgerows, where the piles of soil they excavate, often amounting to several tonnes, are soon colonised by stands of elder which can serve as a good indicator of the presence of badgers.

Bullfinches are quintessential hedgerow birds. Although shy and retiring, they can sometimes be seen gorging on berries in late summer and autumn.

The stoat is one of the more fearsome hedgerow predators. It can be distinguished from the smaller weasel by the black tip to its tail.

Chapter five
❖
Investigating walls, banks and dykes

INVESTIGATING WALLS, BANKS AND DYKES

When we consider field boundaries we usually think, first and foremost, of hedges. But walls, banks and water-filled dykes are also vital components of our landscape heritage: they are historical monuments as well as important wildlife habitats. Some are many hundreds of years old, and have much to say about the ways in which our ancestors developed and managed their environment.

Wall Construction

Methods of drystone walling vary considerably from district to district, but the majority of walls, and certainly the most sophisticated and durable, share a number of basic features. They are generally over 1.3m in height (the standard height in most districts is 4ft 6in, or 1.4m): most determined sheep can clear anything lower. They normally consist of two faces of large stones – the building stones, often referred to as the double – which taper towards the top. The cavity between is filled with small stones (the fill or hearting). At intervals longer stones, called throughstones or throughs, pass through the width of the wall and tie the two sides together.

Drystone walls are typically constructed with two faces of large stones, tapering towards the top and separated by a cavity containing small stones. The two faces are held together by transverse throughstones, and are surmounted by a line of copestones.

A drystone wall on the Sherborne estate in Gloucestershire, with thin slabs of limestone and large copestones.

Most field walls are raised on a foundation of large, square stones, placed in a double line. These are laid level to form a base of between 70-80cm in width, generally excavated around 15cm into the soil, with the flattest side of the stone laid downwards to provide maximum stability for the structure. The stones forming the two sides of the wall are laid above, with those in each course overlapping the joints in the courses below. The stones are usually placed with their long axis at right angles to, ie they are laid into, the wall. They are normally placed level, or else slope slightly downwards away from the wall, in order to prevent the penetration of rainwater. The larger stones are generally used in the lower courses, slightly smaller ones towards the top.

Ideally, the throughstones are spaced at intervals of around a metre, although sometimes more widely if sufficient stones of the required shape – thin, and long enough to pass right through the wall – are unavailable. Often they are placed half-way up the wall, but in some districts two or even three lines are employed. The stones are usually selected so that they do not project more than about 5cm on either side, for if they stick out any further animals will rub against them and weaken the wall.

The large, dark angular stones of millstone grit make for stable but sombre walls. This typical example is in the Pennines, near Whaley Bridge.

The wall tapers gradually to a width of around 38cm, and is then levelled and surmounted by the copestones, which are fairly thin and flat stones placed upright and tightly together. These help bind the two sides of the wall together and also provide weight which helps keep the entire structure in place. The wall is erected in stages, with the central fill being put in place as the wall rises, and with each stone being carefully placed so that it touches the one beside it.

A typical Pennine wall of carboniferous limestone, near Little Longstones in the Derbyshire Pennines. The small, light-coloured stones are laid without obvious courses.

In some districts walls have to climb impossibly steep slopes, and this took (and takes) particular skill. If a slope is less than 15°, the courses in the outer faces are laid parallel to the ground, so that they follow the natural undulations of the ground surface. But where the angle is steeper, the foundations are usually formed into a series of short steps, and the courses are laid parallel to these, rather than to the ground surface. In these circumstances the copestones are often angled slightly, so that they lean uphill.

Varieties of Wall

Although most walls in Britain are constructed according to the basic pattern described above, there are – as with hedges – many local variations. In part these reflect the character of raw materials: walls were made of local stone, and different kinds of stone come in different shapes and sizes, and are cut and shaped with widely varying degrees of ease, encouraging the development of different walling traditions. But they also reflect the fact that the 'normal' form of construction only developed in the course of the eighteenth century. Earlier walls were built in rather different ways.

Comparatively little work has yet been published on the development of stone walls, and the best study remains that carried out by the archaeologist Richard Hodges and his team at Roystone Grange in Derbyshire. Here, three main forms of construction were identified, each related to a particular phase of the area's enclosure history. Walls of eighteenth- and nineteenth-century origin, and those substantially reconstructed in this period, display all the classic features just described. They are built of fairly regular blocks of limestone, apparently (in most cases) deliberately quarried, and were set in shallow trenches made by removing the turf and topsoil. Walls of sixteenth- or seventeenth-century date, in contrast, were raised on a base of large dolomite boulders set directly on the ground surface. They also have two faces of large stones, sloping inwards, but lack a dividing void filled with hearting. Moreover, while they are sometimes capped by trimmed copestones, they are often topped by a line of large boulders. Throughstones are absent or sporadic: all in all, these are less accomplished structures. Only a few walls of medieval date survive at Roystone Grange, and these have often been extensively rebuilt, but all consist of a single row of very large boulders laid directly on the ground surface, with upper courses comprising a single thickness of rough boulders.

A neat, regular limestone wall near North Rauceby in Lincolnshire. The drystone walls of the narrow 'Heath' district to the south of Lincoln seem curiously out of place in this intensively arable countryside.

This general pattern of development – from massive and irregular construction, through walls with two faces but no throughs or hearting, to walls of 'normal' construction – is apparent elsewhere, although with important variations. In the Yorkshire Dales, for example, the oldest walls consist of very large stones – often almost one metre across – arranged in a rough double wall: the space between is filled with stones of all shapes and sizes, cleared from the area around. Such walls have wide bases – sometimes as much as 2.5m – and narrow markedly towards the top, so that they resemble in profile a blunt pyramid.

Age is thus one important explanation of variations in construction. Geology is another. Walls were constructed of stone picked up from the adjacent land, or quarried in the immediate vicinity. Indeed, very subtle and detailed changes in geology can be reflected in patterns of walling, as for example in Wensleydale, where the alternating limestones and grits give rise to marked dark and light strips in the walls as they climb the fells. These detailed changes are a microcosm of the wider distinction in Pennine walls, between those made of millstone grit and the

Hedges and walls

various rocks of the coal measures, and those constructed of limestone. The former stones are dark brown in colour and are the main walling material in North Yorkshire and in much of Derbyshire. The gritstone comes in rough blocks and flags, which can be used to build fairly regular walls; the coal measures give rise to still neater, more evenly-coursed walls. There are many suitable throughstones, and the rough surface of the stones provides good adhesion so that walls are often quite narrow and do not taper markedly towards the top. Their rather elegant appearance is sometimes enhanced by the use of carefully shaped copestones, especially in villages. These walls contrast sharply with those constructed of carboniferous limestone, which can often be found in close proximity. These walls are much lighter in colour, almost white in some places, and they are built of rather smaller and more irregular stones, laid without clear lines or courses. Moreover, in some districts suitable throughstones are in short supply and so the walls tend to have wide bases to give extra stability.

In the Lake District there is even greater variety in the style and appearance of field walls, reflecting a more diverse geology. In the south-east of the region, in the area between Kendal and Lancaster, carboniferous limestone is again the dominant walling material, and the walls are in consequence light in colour and slightly irregular in appearance. Further north, towards Ambleside, the Silurian slates and shales come as regular, thin blocks which form neat, regular walls. Further north still, in the district extending from Ambleside to Keswick, a wide variety of igneous rocks are found, hard and difficult to cut to shape. As a result, the walls have an irregular appearance, containing stones which vary greatly in size, and often have very large volcanic boulders in their lower courses. Some huge walls, 2.5m or more in height, occur sporadically. Moving still further north, onto the fringes of Lakeland, walls once again become neat and tidy in appearance, for here they are constructed of the dark-coloured or greenish Skiddaw slates, which cleave easily into regular blocks, again allowing for regular courses. However, the diversity of Lakeland walling does not end there; further stretches of limestone wall occur in north-west Cumbria, rusty-red walls of New Red Sandstone are found in the Eden valley, while around Coniston, Hawkshead and Ambleside blocks of rough slate are sometimes fixed upright, in a line, to form a crude fence, sometimes with a hedge planted behind it.

Wasdale in Cumbria boasts some fine examples of substantial field walls constructed of volcanic rock.

An unusual stone wall, made of individual slabs set upright, at Kelmscott in Oxfordshire.

Very different are the Jurassic limestone walls found in the Cotswolds. Here the better building stone, from the thicker oolitic beds, is quarried as freestones, for the construction of houses and farm buildings. The poorer, rather shelly limestone, found closer to the surface, is used for field walls. It provides stones of an attractive golden colour, but they are small, light, and rather soft. There is a shortage of stones large enough for throughs and copestones. Partly because of this Cotswold walls tend to be lower than those found in the upland areas of Britain, often being less than 4ft (c.1.3m) high, and the copestones or 'combers' are sporadically mortared.

In some cases, especially in villages, the softness of the stones allows the wallers to trim them into shape, producing very neat and regular walls. Out in the fields, in contrast, the stones are generally less carefully prepared and are often laid with their length along the wall, rather than at right angles to it, the normal practice. Cotswold wallers take particular care to slope the surfaces of the stones away from the wall in order to prevent penetration by water, for the soft stone is particularly susceptible to frost damage. Similar but rougher stones produce yet more irregular walls in the heath district of Lincolnshire, many now in a very poor state of repair, and sporadically in parts of Northamptonshire and Leicestershire, where they are sometimes combined with thorn hedges. At the other end of the Jurassic limestone belt, on the Isle of Purbeck, the limestone is lighter and coarser, the walls bright but of rather crude construction.

In the Mendips carboniferous limestone again provides small and irregular stones, although here the walls tend to be less broad and less tapering than in the north of England. To overcome the paucity of stones suitable for throughs, stones are used which extend some two-thirds of the way across the wall. Here, too, it is the custom to lay the stones with their length along, rather than set into, the wall.

In Wales, diverse geology has engendered a bewildering variety of walling techniques. In the south of the country, on the Gower Peninsula, limestone and sandstone are both employed, the former producing the most regular and durable walls. Further north, in the Welsh coalfields, various forms of gritstone give rise to walls broadly similar in appearance and construction to those found in the

A Cornish hedge at Lower Porthmeor in West Penwith. The massive boulders are a characteristic feature.

millstone grit areas of the Pennines. In Gwynedd, however, a wide range of volcanic rocks is employed for walling. In most cases these are hard and difficult to trim, giving rise to rough grey walls. Some are substantial structures, more than two metres high and a metre and a half thick, and evidently created primarily to clear stone from the fields. The use of throughstones is often limited and the copestones are sometimes omitted altogether, the sheer bulk of the wall being considered enough to keep it upright. Where slate is quarried, especially around Llanberis and Bethesda, the waste is extensively used to produce more regular and accomplished walls. In some places the two kinds of stone are combined, with unusual effects, and fences made with lines of vertical strips of slate, wired together, are also occasionally encountered, most being of relatively recent date.

Drystone walls are also found in parts of Cornwall and Devon, mainly around the edges of the moors – especially Dartmoor and Bodmin. Where slate is available, neat well-coursed walls are the norm. But hard granite and other igneous rocks give rise to rather crude walls, laid in rough horizontal courses and sometimes topped with turf. Often massive stones form the base of such walls, particularly massive in the more ancient examples. The walls on Dartmoor, constructed of stones cleared from the surrounding ground, often have lines of gorse growing along them, bright with yellow flowers in season. In most cases the gorse has colonised naturally, finding a haven from grazing sheep, but in places it may have been deliberately planted in order to provide a more secure barrier.

Special Features of Walls

Traditional field walls, especially those in northern districts of England, incorporate a number of specially built features, which again display various degrees of local variation. The most common of these is a wall-head, check or gate-end, where a wall comes to a free-standing end. This is most usually found where a gap needs to be left for a gate but can also serve other purposes. For example, two wall-heads next to each other were sometimes used to show where responsibility for the upkeep of a communal wall changed from one proprietor to another. Large stones are built up, layer on layer, in such a way that those running across the width of the wall alternate with those extending back into it. The stones need to be particularly regular in shape, and these are not always easy to find in areas where the majority of walling stones are small and irregular, as in the carboniferous limestone districts of the Pennines.

Wall-heads cannot easily be used to hang a gate, and for this gateposts were used. These might be of wood, of unshaped local stone, of shaped local stone or of shaped stone brought in from another area. The last only really became common practice from the middle decades of the nineteenth century, when the spread of the rail network facilitated the movement of such bulky items. In the carboniferous limestone White Peak area of the south Pennines, for example, gateposts of shaped millstone grit were widely used after the middle years of the nineteenth century, but seldom before.

Hedges and walls

Another common feature of drystone walls is the creep-hole, cripple, lunky, hogg-hole or smoot, used to allow sheep to move freely from one field to the next. This is a small hole at the base of the wall, with a particularly strong stone as a lintel: when required, the hole could be closed easily by a large stone or by wooden planks. A similar construction is often found where walls run across small streams. There are many variations in the detail of such features, some again a function of the character of raw materials and some related to age.

Richard Hodges and his team at Roystone Grange were able to distinguish sheep creeps of medieval and early post-medieval origin from those of the eighteenth and nineteenth centuries. The former have single dolomite boulders for jambs and lintel and a roughly square opening, whereas in the latter the lintel is a large slab of carboniferous limestone, the jambs are constructed from a number of fairly regularly shaped stones and the opening is roughly twice as high as it is wide.

Stiles are necessary where walls are crossed by footpaths. They come in a number of forms, but two in particular are often encountered. 'Squeeze stiles' are V-shaped openings in the upper part of the wall, or (more rarely) narrow openings running the full height of the wall. The second, more usual, type consists of extra long stones which project out from the wall to provide footholds so walkers can climb over. Once again, local variations in the frequency of different forms are related to the nature of the available materials, to the age of the walls in question, and to local farming practices. Hodges and his team suggested that stiles consisting of stones projecting out from the wall were of seventeenth-century date, while those comprising openings in the wall dated from the enclosure period of the late eighteenth and early nineteenth century. But the precise character of chronological variation varies from area to area, and would repay further local study. Perhaps the main factor was the kind of animal grazed in the adjacent fields: squeeze stiles would act as a barrier to cattle but not, in many cases, to sheep.

Dating Walls

Attributing a date to field walls is not easy, not least because the date of the boundary is not necessarily the same as that of the wall which runs along it. The wall might have been completely reconstructed at some later date on existing foundations, a practice which seems to have been particularly common in the later eighteenth and nineteenth centuries. In general, as we have seen, earlier walls are massive and irregular in construction, often with wide bases, while later walls conform more to the 'normal' type, with regular throughs and copestones. But limitations imposed by local materials ensure that there is no inevitable

Old walls are home to a variety of plant life, including pennywort and *Rhizocarpon* lichens.

progression, and many quite late walls – including some of those created by parliamentary enclosure – may display 'early' characteristics. In particular, walls with massive boulders at the base continued to be erected around intakes on the North York Moors well into the eighteenth century, while in some of the limestone areas of the Pennines even parliamentary enclosure walls can have quite wide, spreading bases, because of the lack of suitable throughstones. Certain other features of construction are worth looking out for. In particular, where all or most of the walls within a given area conform to the same dimensions, in terms of height, width, and batter, they are usually the result of a parliamentary enclosure. As we have seen, enclosure awards often laid down detailed specifications for the walls surrounding the new allotments.

As with hedges, some attempts have been made to use botanical indicators as evidence of age. Lichens are an important feature of many walls, and will vary according to location and rock type. Thus on acid rocks in the highlands, map lichens – yellow species of *Rhizocarpon* – are usually conspicuous; while on limestone walls the pits of *Verrucaria baldensis* are a notable feature. Different species of lichen grow and colonise at different rates, and geologists have for many years used the technique of 'lichonometry' to estimate the date of natural rock exposures. Similar methods have been used to date gravestones, and some walls – as at Bradgate Park in Leicestershire. However, there are many problems in applying this technique to field walls, not least the fact that identification of lichens is a difficult and expert task. Walls are frequently repaired and rebuilt, and individual stones might have been put in place long after the wall's construction. Conversely, where the walling material was derived from land clearance, lichens might already have accumulated on individual stones before construction. Moreover, lichens are very sensitive to air pollution, effectively ruling out the technique in the more industrial parts of Britain.

A good guide to the age of a wall is provided by the shape and layout of the boundary it defines, just as it is with hedges. Boundaries of medieval or early post-medieval date, for example, those marking the perimeters or subdivisions of vaccaries, the bounds of arable fields or the head dykes separating them from the open moors, the edges of the stinted grounds or cow pastures carved out of the commons in the sixteenth or seventeenth centuries, intakes from the waste, or boundaries between the common grazing of neighbouring communities, tend to run in sweeping or irregular lines. They are seldom, if ever, dead straight and often take advantage of natural features in the landscape, such as outcrops of rock or individual boulders too large to move. Such boundaries can usually be distinguished without too much difficulty from those created, usually during the fifteenth, sixteenth and seventeenth centuries, by the piecemeal enclosure of open fields, for these have a sinuous, slightly curving appearance (sometimes in the form of a 'reversed S'), replicating that of the old open-field strips. These in turn can be distinguished from the boundaries created by parliamentary enclosure, or by changes to existing boundary patterns effected by large landed estates, in the eighteenth and nineteenth centuries. Such boundaries normally run in dead straight, surveyed lines, sometimes up or obliquely across surprisingly steep slopes – as if the surveyors who divided up open moors were familiar with the landscape on a map, but had never really got to know it on the ground.

Many thousands of miles of drystone walls still survive in England and Wales, but many are in poor condition, tumbled and gappy. They have often become superfluous to requirements. Over the years farms have increased in size through the amalgamation of holdings, and small walled fields are unsuitable for tractors, hay cutters and other modern machinery. Where stock-proof barriers are required these are often now constructed of barbed wire or sheep netting, and even where walls survive they frequently have such fences erected along or above them. The present condition of field walls in any area is, perhaps not surprisingly, related to the type of stone used to build them. Thus in the Peak District walls constructed of millstone grit are generally in rather better condition than those built from the thin and irregular fragments of carboniferous limestone. Recent decades have seen a welcome resurgence of interest in walls and walling and, while many boundaries continue to decay, the importance of walls as defining elements of many upland landscapes is more widely recognised than ever before.

Wall Wildlife

Walls are also of great significance as wildlife habitats. A wide range of lichens, mosses and ferns can be found on them, together with many members of the saxifrage and stonecrop families. Not surprisingly, the character and variety of plants present is related to the geology of the constituent rocks, with acid rocks like gritstone carrying a different, and usually sparser, flora than limestone. On walls in the Mendips or the southern Pennines, for example, the orange lichens of the genus *Calopaca* are frequently encountered; spleenworts and ferns are widespread, including such rare species as the limestone polypody and the brittle bladder fern.

The south-facing elevations of drystone walls in particular can support xerophytic plants, such as rue-leaved saxifrage and wall whitlow grass, as well as exotic escapees from village and suburban gardens, like red valerian, Oxford

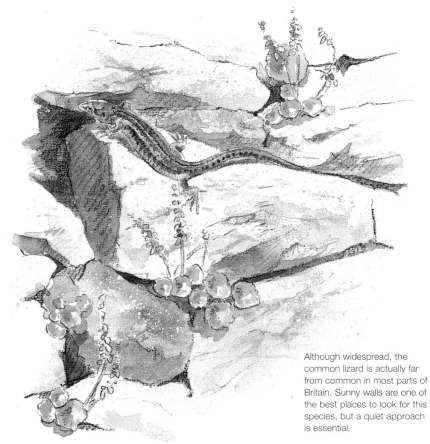

Although widespread, the common lizard is actually far from common in most parts of Britain. Sunny walls are one of the best places to look for this species, but a quiet approach is essential.

ragwort and ivy-leaved toadflax. These last three species are all introductions to Britain from the Mediterranean, and are now thoroughly naturalised and widespread. The micro-climate offered by sheltered walls suits them well, as they thrive in hot sun and are able to survive long periods of drought.

The ivy-leaved toadflax is a real wall specialist and in Britain is hardly known away from man-made habitats. It has evolved a reproductive technique that is particularly well-suited to wall life: its small flowers initially grow upwards and outwards on long stems, but when pollination has been achieved the growing stalk turns away from the sun, back towards the wall, eventually pointing its seedhead into a convenient nook or cranny. The seeds then drop off and germinate in the sheltered conditions between the stones; in this way the plant is able to keep seed wastage to a minimum and can effectively colonise walls vertically as well as laterally.

Walls can also harbour a rich and diverse fauna. They are especially valuable for spiders, including the zebra spider, whose jumping ability makes it one of the most effective predators in the wall environment. Leaping from stone to stone in search of prey, this species shows considerable ingenuity. When faced with an apparently insuperable gap, the spider will feed a line of silken thread into the wind, waiting for the thread to snag on a nearby stone and thereby provide a means for it to cross. Walls are also home to a wide range of snails. These spend the winter hibernating deep within the wall before emerging in spring in search of new plant growth. They generally feed at night and can travel surprising distances before returning to the same crack or crevice by dawn. This sense of territory, together with a rather surprising longevity, means that an individual snail can frequent the same patch of wall for several years.

Insect life abounds in and on walls, and includes species that have become particularly associated with the habitat, such as the wall brown butterfly. The warm, south-facing stones so favoured by this declining insect are also the place to look for reptiles such as the common lizard and slow-worm. Mice, field voles and bank voles also find food and shelter among the stones and are in turn preyed upon by weasels, stoats and – in Wales and the Marches – polecats. Birdlife can be interesting too, with upwards of thirty different species having been recorded nesting or roosting in walls. These include familiar species such as wren, blue tit, great tit, house sparrow and pied wagtail, as well as the less common nuthatch and redstart. In upland areas the

FACING PAGE Walls provide habitat for a wide variety of plants and other wildlife. The wheatear and red valerian are both common species in many parts of Britain.

Snails are most common on limestone walls, the calcium in the stone being vital for their shell-building.

wheatear is closely associated with walls, its vivid white rump a familiar sight as it flits along the wall-tops. For certain raptors, including merlin and little owl, walls also serve as important vantage points from which to survey territory and seek out prey (and the owl will happily breed in them also). Some decidedly unlikely species also take to using walls – in favoured areas near the coast, the ocean-going storm petrel will even nest within them.

The little owl was introduced to Britain from mainland Europe in the nineteenth century. Filling an apparently vacant niche, it soon spread and is now fairly common across England and Wales.

Banks

In several parts of England and Wales fields are enclosed not by walls or hedges but by substantial banks of earth, mixed or faced, to varying degrees, with stones. The most famous and distinctive are the so-called 'hedges' of western Devon and Cornwall. These come in a variety of forms, with research by the Cornish Archaeological Unit suggesting that around twelve main types can be identified. Some are massive features, with foundations of huge boulders cleared from the fields, like those – mainly of prehistoric origin – found in West Penwith and on the Land's End peninsula. Others are lower, lack the massive boulders, and contain

A Cornish 'hedge' is really a bank composed of two faces of stones, often very large, with an infill of earth. Shrubs – both opportunistic and planted – often adorn the top.

The churchyard at Tintagel is surrounded by a Cornish hedge, faced with slate.

more earth than rock. Some, especially those in more exposed coastal locations, have few shrubs growing along the top – often no more than the occasional gorse or hawthorn; some have none at all, the top being covered with turf and a mix of herbaceous vegetation. But others, especially those along roadsides and in more sheltered locations, carry an almost continuous line of shrubs, reducing the contrast with a 'normal' hedge. Gorse, bramble, heathers (both *Erica* spp. and *Calluna vulgaris)*, and bilberry are all frequent, together with hawthorn and blackthorn. Hedgerow trees of oak, ash and sycamore are found in more sheltered inland areas. Most such hedges also carry a rich herb flora, which varies with geology and climate. In exposed coastal areas thrift, kidney vetch and sea campion are prominent; in inland areas, red campion, honeysuckle and foxglove.

In the majority of cases the width of the 'hedge' at its base is roughly the same as its height. The bank tapers to about half this width at the top. Most are faced with stone, and in some cases the two faces are slightly concave, probably because the builders over-compensated for the bulging that can occur in the first years of the wall's existence, as the earth fill gradually settled down and pushed the lower courses of the facing outwards. The precise method of laying the facing stones varies greatly: this is partly a function of age and local tradition, partly the result of the character of the available stone. Where thin slabs of material dominate, the stones are often laid in herringbone fashion, like the fine examples to be seen on National Trust property near Polzeath in Cornwall. Even where the majority of stones are laid in more conventional horizontal fashion, the top two courses are often laid herringbone-style, for this allows the roots of the turf or vegetation to

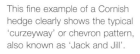

This fine example of a Cornish hedge clearly shows the typical 'curzeyway' or chevron pattern, also known as 'Jack and Jill'.

spread rapidly downwards and thus provide binding and additional stability for the wall. Such considerations probably explain why, in marked contrast to the practice when erecting more conventional drystone walls, the stones are often laid so that they tilt back towards the centre of the wall. As the builders wanted vegetation to colonise the 'hedge', water had to be encouraged to reach its depths. But there are many local styles and variations. In some parts of Devon, for example, a technique called 'chip and block' is often employed. The gaps between the large stones are filled by smaller chips, often (on estate hedges in particular) producing a very smooth finish.

Banks of earth and stone, but often lacking such a carefully constructed stone face, can be seen in a number of other western districts of England, notably on the coastal plain of Cumbria, where they are called 'kefts'. These are often topped by a continuous line of hawthorn or gorse. Particularly fine examples can be seen above the River Eden in Egremont. On the Isle of Man, according to local tradition, the gorse was added to the banks some time after their construction, to provide a more effective barrier to cattle. It is noteworthy that the species of gorse most frequently employed is not that native to the island, western gorse, *Ulex gallii,* but rather common gorse, *Ulex europaeus* imported from the mainland. Gorse, either planted or gradually colonising, is a notable feature of other banks in western Britain.

In parts of west Wales and the Llŷn peninsula earth banks are also the normal form of enclosure. Many have a hedge growing along the top and most are faced with stone, sometimes set vertically, sometimes in herringbone fashion. Earth banks are also common in the north of Ireland. The majority of field boundaries here, as we have seen, are of relatively recent origin, the consequence of the 'improvements' carried out by landowners during the eighteenth and nineteenth centuries: and most of these are composed of hedges dominated by hawthorn. In some areas of Ulster walls of irregularly piled slabs with tall upright stones placed at intervals are found. But many boundaries, both early and late, comprise some form of earth bank, topped with gorse and bramble. Many have one sloping and one vertical side, both faced with stone: thorns are planted between the stones near the base of the wall to provide an additional barrier.

Substantial banks of stone and earth are thus essentially a feature of western Britain, and especially of coastal areas where high winds militate against the successful establishment of hedges of more normal form. However, less imposing earth banks can be found in other places. In parts of the New Forest fields are often defined by low sod banks, sometimes combined with gorse; while in the windswept coastal areas of north-east Norfolk low banks, with occasional shrubs and oak trees, are sometimes found. In 1787 William Marshall noted these 'bare mud walls' as a particular feature of the district. Most were probably originally planted as hedges, which have failed to flourish in the harsh conditions of the Norfolk coast.

FACING PAGE Drainage ditches are of immense importance as wildlife habitats. The Gwent Levels beside the Bristol Channel contain many fine examples.

In terms of wildlife, earth banks can support interesting plant communities and are home to familiar species such as teasel, field scabious and cowslip, as well as to certain types of orchid, most commonly bee, pyramidal and twayblade, and many varieties of grass. They can also be important for specialist insect species, including ants, bees and wasps. South-facing exposed banks warm up readily in the sun and so provide ideal conditions for nesting. Light, friable soils make for particularly easy digging, and often the nesting holes of hundreds of mining bees can be seen clustered together on a favoured bank. Burrowing beetles, notably the minotaur, also enjoy these conditions, particularly so where mammals such as rabbits or sheep are present – the minotaur gathers their dung pellets and stores them in underground chambers as food for its larvae. Butterflies such as the small copper, common blue and, more locally, the small blue, will also thrive on earth banks if the requisite foodplants are present.

Dykes and Ditches

In many wetland areas of England and Wales, hedges, walls and banks are replaced by dykes or 'rhines' as the principal form of field boundary: that is,

Ditches were often created for reasons other than drainage. This example – Docwra's Ditch at Dunwich in Suffolk – was actually dug as a reservoir and firebreak (the surrounding heathland is very dry).

permanently water-filled ditches that lack an accompanying bank. Dykes drain the land and form a complex interconnected network which then ultimately discharges the water into a nearby river or the sea. The water sometimes passes through the 'wall' or embankment protecting the marsh from inundation via a simple 'flap sluice', held closed by water pressure at high tide: this was the normal method in medieval times. More usually the water is pumped over the bank, originally by windmills (adopted on a large scale in the seventeenth and eighteenth centuries); later by steam or diesel pump; and nowadays by pumps powered by electricity.

Dykes are not only drains. They also function as 'watery fences', serving to demarcate property and preventing livestock from straying from one section of marsh to another. They also provide stock with drinking water. Where marshes were used for grazing, the traditional practice was to put animals out in the spring after the 'freeboard', or water level, in the dykes had been set to a height of about 2ft (60cm). In the late autumn, when the animals were taken off again, the sluices were adjusted so that the water level fell, ready for the dykes to receive the increased winter run-off from the surrounding high ground.

Wetlands take a wide variety of forms, have very diverse histories and display a bewildering variety of boundary patterns. The oldest dykes are found in coastal and estuarine marshes, areas of silt and clay, rather than in areas of fen associated with peat soils. The former are rich and fertile lands which

were usually reclaimed at an early date and, even in the Middle Ages, were often ploughed as arable land. Peat fens, in contrast, which are often found adjacent to (and inland from) silt marshes, were mainly used as great wet commons until enclosure and reclamation in the seventeenth, eighteenth and nineteenth centuries.

Individual dykes are difficult to date – their form is often reconstituted, and their biological clock reset, by regular 'slubbing out' – but some on the Wentlooge Levels in South Wales have been shown to be of Roman origin. A number of the close-set, parallel dykes here continue beyond the modern sea wall as lines of darker soil in the tidal mudflats, for they have been truncated by later coastal change. When these relict features were excavated they produced fresh, un-abraded sherds of Roman pottery. Roman colonisation was widespread in other coastal marshes in England and Wales but changes in sea level in early Saxon times normally led to abandonment, and subsequent re-colonisation, so that the earliest dykes are usually of late Saxon or medieval origin.

On the silt 'Marshland' of Norfolk and Lincolnshire the earliest fields, laid out around villages established in later Saxon times on the higher ground towards the coast, are rather irregular in shape. Those created by subsequent reclamation inland during the eleventh, twelfth and thirteenth centuries took the form of long, narrow strips, seldom more than 20m wide and often considerably narrower, yet in extreme cases as much as a mile or so in length. Many of the dykes around them survive, although others have been filled in to facilitate modern land use, and the distinctive strip pattern has been greatly simplified. Different again was the medieval dyke pattern on Romney Marsh, where a myriad of tiny dyked fields was created, most of less than 1.5 hectares, in part by adapting the natural salt-marsh drainage channels. Here, too, subsequent changes have served to simplify the pattern, although in places traces of the original, dense network survive in earthwork form. Similarly, on the Halvergate Marshes in east Norfolk most of the dykes have the curving, meandering form characteristic of such relict natural watercourses. They are interspersed with much straighter dykes, evidence of attempts to improve drainage in post-medieval times.

Such ruler-straight dykes completely dominate the landscape in most areas of peat soils, where open common fens generally survived until post-medieval times. Before this, fens had been comparatively wild, undrained areas, grazed to some extent but mainly cut, for peat, fodder and thatching materials. Yet even these comparatively recent dyke patterns are often full of interest. The great peat fens of East Anglia have a superficial monotony – hectare upon hectare of rich arable land stretching to the horizon – but closer inspection reveals numerous different blocks of fields orientated in different directions. Some represent allotments given to investors and wealthy landowners following the drainage schemes of the seventeenth century, others constitute areas enclosed and divided at various times in the eighteenth and nineteenth centuries. Similar patterns of straight dykes enclosing precise and geometric fields can be seen in the inland peat areas of the

Somerset Levels – in the valleys of the rivers Brue, Axe and Cary, and on the King's Sedgemoor, areas which were mainly reclaimed in the eighteenth and nineteenth centuries. Here, however, drainage was followed by grazing rather than by ploughing and the area was divided into fairly small dyked fields, some of which were used for peat digging.

Many of the Somerset dykes are lined with pollarded willows, and these are a noticeable feature of boundaries in other wetlands: most noticeably, the close-set lines which grow beside the roads crossing the Halvergate Marshes. Willows were regularly cut and the material used for making baskets, eel traps, hurdles and much else. But in some cases this was a secondary benefit, and the main purpose of the trees was to reinforce the bank and, in the case of roads, provide stability for the road surface. Pollarding kept the trees low (very low in the case of the Halvergate examples) so that they would not be brought down by the wind. It also allowed the trees to be spaced very narrowly, and this provided the maximum amount of root structure to bind the ditch banks, or 'hold the shoulder' of the road.

A typical drainage ditch on the Somerset Levels, bordered by pollarded willows. These trees provide valuable roosting and nesting sites for both barn and little owl.

Wildlife in Dykes

Dykes have a considerable historical significance, and some of those in areas of drained coastal marsh are a thousand years old. But they are also vital wildlife refuges, particularly in areas where drainage and reclamation have reduced the amount of available wetland habitat. Water pollution and the impact of increased boat traffic on main rivers have also served to curtail the opportunities for certain water-loving plants. Species such as marsh marigold, brooklime, watercress, water violet, water soldier and marsh horsetail are all characteristic of dykes, so long as the water remains in healthy condition. A wealth of invertebrates also occupies dykes and drains with good water quality, and can be divided broadly between the species that live on the surface of the water, those that live below the surface (including on the bed of the dyke), and those that inhabit the waterside vegetation. Many species actually take advantage of all three categories of habitat at different stages in their life cycle – caddisflies, for example, lay their eggs on the water surface, with their larval and pupal forms developing underwater and the adult insects later emerging to gather in large numbers over the water and adjacent vegetation.

Pollution-free dykes and ditches can be teeming with invertebrate life. Whirligig beetles and water boatmen are typical inhabitants.

Typical surface life includes pond skaters, whirligig beetles (although these can also dive in search of food) and springtails, one of the most primitive forms of insect. Living below the surface in areas of more open water are predators such as the water boatman, water scorpion and the great diving beetle, one of the largest (and most voracious) aquatic insects. The bed of a dyke, rich in organic silt and debris, is where the larval stages of many species take place. Many can even cope in the deoxygenated environment of stagnant ditches.

A particularly interesting invertebrate species found in dykes is the swan mussel, the largest and most common of the 28 varieties of freshwater mussel found in Britain. Like the various types of aquatic and amphibious snails which also thrive in ditches and dykes, the chief threats to the swan mussel are drought and overzealous dredging. Pollution poses a further threat to all forms of life in dykes, especially in areas of intensive agriculture where the leaching of pesticides and other agro-chemicals into nearby watercourses can have a serious impact.

Hedges and walls

Once common along rivers and dykes, the water vole is now a rare sight. Although the spread of the mink has been blamed for this slump in numbers, it is now becoming clear that pollution and habitat destruction are the principal causes of the vole's decline.

A voracious hunter, the American mink originally escaped from fur farms and now lives ferally across much of Britain. However, the recent resurgence in otter numbers appears to have checked the mink's advance.

Dragonflies and damselflies can be significant along dykes with well-established plant life. The banded demoiselle is one of the most frequently encountered species, but it is not unusual to find up to ten varieties of damselfly and dragonfly along dykes in certain parts of England. They compete for airspace with a myriad of other airborne creatures, including mayflies, stoneflies and dance flies, which in summer gather in huge clouds above the water, the males and females jostling in crazy whirls as a prelude to mating.

This varied plant and insect life in turn attracts a range of bird species. Both sedge and reed warbler will breed along dykes if there is adequate vegetation (*Phragmites* reed being essential for the latter species), as will reed bunting. The opportunities offered by dykes can provide a niche for species that might otherwise not be present in what are often intensively farmed environments. For example, herons will regularly commute to them in order to hunt amphibians and fish, with mute swan, moorhen, coot and, more elusively, water rail also often present in such habitat. Along with the ubiquitous three-spined stickleback – hardly worth the attention of any self-respecting heron – eels are the most frequent fish in many dykes. They penetrate well inland and are a major source of food for Britain's growing heron population.

The relative usefulness of dyke habitats to birds is a complex subject and varies according to the particular requirements of individual species; for example, watercourses in which the water surface is obscured in summer by a seemingly impenetrable 'lawn' of duckweed are clearly of less value to birds such as the kingfisher, which depends heavily on visibility to find its prey. Even so, where dykes are in good condition – and their banks not too scrupulously 'managed' – birdlife can be both prolific and surprisingly diverse.

Among mammals, the water vole was once a regular denizen of dykes but is now much declined and even absent from many areas. An expert swimmer, the vole is equally at home on dry land and in the winter months shows a preference for drier conditions, sometimes burrowing its tunnel systems away from the water's edge and under adjacent fields. In this respect drainage dykes can offer good habitat for this species. Part of the vole's demise is attributed to the spread in recent years of the introduced and highly versatile American mink, which has moved into waterside habitats with both speed and enthusiasm. Equally at home on land or in the water, mink feed mostly on fish during the winter months, but in spring and summer they turn their attentions to small mammals, amphibians, birds and eggs, often with devastating results.

The cream and rather sickly scented flowers of meadowsweet and the feathery brushes of water horsetail are a common sight along ditches in summer. Meadowsweet was originally used to flavour mead, and horsetails to scour pans (their rough stems and leaves acting like wire wool).

Chapter six

❖

The future

National Trust volunteers repairing a Cornish hedge. Interest in the care and management of boundary features has certainly increased in recent years, but practical improvements are still patchy and some worrying threats remain.

PREVIOUS PAGE The rolling countryside around Strangford Lough in Northern Ireland has retained many of its hedges.

THE FUTURE

Over the last decade or so the tide of destruction has turned and the rate of field boundary removal has slowed noticeably. This is due in part to changes in government policies, but these are themselves largely a reaction to wider changes in social attitudes. Worries about food shortages now seem far away, and have been replaced by widespread concerns about the degradation of the natural environment. Social changes have also played a part. More people than ever before visit the countryside, and more and more vocal, middle-class residents settle there. For every Brian Aldridge there is now, in many areas, a crowd of Linda Snells. Farmers and landowners are not immune to these wider changes in attitudes and, encouraged in particular by the activities of the Farming and Wildlife Advisory Groups, began to re-plant hedges during the 1980s. They were spurred on by a genuine concern for the quality of the environment, but an active interest in game conservation and worries about the decline of birds like the grey partridge also played their part.

The rate of replanting increased markedly during the 1990s and, although the grubbing out of hedgerows still continued, gains and losses began to move into equilibrium. Various government schemes, intended to re-direct resources away from over-production towards nature conservation, offered financial incentives to help protect and even replenish hedgerows. Foremost among these have been Countryside Stewardship and the Environmentally Sensitive Areas scheme, whereby farmers are paid to enhance the environment rather than simply produce food. Furthermore, mounting public concern, especially as hedge removal seemed to be continuing at an alarming rate in some areas of England, encouraged the Conservative administration under John Major to bow to pressure and bring a stick, as well as carrots, to the problem. In 1997, as one of their last acts, the Conservative government brought in the Hedgerow Protection Act, and this was embraced by the incoming Labour administration.

The law has been under constant review and is by no means a perfect piece of legislation. It is oddly phrased, in places ambiguous, and strictly observed it would probably not provide protection for much more than 40% of existing hedges. Nevertheless, challenges to the act brought by landowners concerning particular hedges have been relatively few. Combined with the subsidies increasingly available for planting and maintaining hedges, the new law has ensured that hedge removal has declined dramatically and, at the time of writing, the length of hedgerows in England is probably increasing for the first time since the early nineteenth century. Moreover, the situation may well improve further in the medium term. Subsidies for

Unkempt field margins and hedge bases are vital for the breeding success of birds like the grey partridge, shown here. Its red-legged cousin – an introduced species – is much more of an open field bird and so potentially less affected by hedge removal.

production are predicted to fall, as the European Union struggles to control the escalating cost of the Common Agricultrual Policy, and as international free trade agreements limit the amount of subsidies states can provide to support food production. Yet at the time of writing the budget for Countryside Stewardship schemes is set to increase, from around £11 million in 1997 to £126 million in 2006/7.

But even in this more favourable climate hedges remain contentious. There are, for example, widely varying opinions about how best to establish new hedges. A 'mulch mat' of some appropriate biodegradable material helps protect the young hedge from being smothered by weeds, but most nurserymen also recommend the use of tree guards or 'spirals' to protect the plants from the attentions of rabbits, hares and deer. The difficulty here is that this produces plants on tall stems and a hedge with a very gappy base, which is useless for wildlife. Drastic cutting, and simple laying, are soon required to thicken it. Ideally, planters should avoid using guards and surround the entire hedge with rabbit-proof netting, but this is a very costly option and so seldom used.

In many ways it is easier to establish new hedges in arable areas, where they are functionless adornments to the landscape, rather than in many livestock areas, where they ought in theory to serve some useful function. Sheep are great devourers of young shrubs, and until a hedge is well grown it needs to be protected with sheep netting, and on both sides if both sides are grazed – again an expensive option. Hedging plants of local origin should be used, but at present there are insufficient supplies in most areas and seed is therefore brought from far afield, often from nurseries in Holland. This is an unfortunate practice, as it dilutes or destroys the genetic uniqueness of local stock.

The craft of hedging has made a notable comeback in recent years. Here a mature hedge is being laid in the traditional way, with upright 'stabbers' held in place with 'binders' or 'hethers'.

The management of both new and existing hedges is also a matter for some debate. Many conservationists advocate plashing, but this will never be practical on a large scale – nor, as we have seen, was it ever the only form of hedgerow management. Most surviving hedges are cut using a mechanical flail or – occasionally – a rotary saw fixed to a tractor. The results look messy and, carried out at the wrong time of the year, or too frequently, can damage the hedge and local wildlife. Cutting in spring and early summer, when birds are nesting, is disastrous but unusual: but cutting in early autumn, when the hedges are laden with haws, nuts and berries, is still distressingly common. Mechanical cutting is the only viable method of maintenance in most circumstances, and drastic coppicing by mechanical methods on a rotation of ten to fifteen years ought, perhaps, to be more widely adopted.

A densely laid hedge is not necessarily ideal for all types of wildlife, and for many birds a high and bushy hedge is better, or a hedge mechanically maintained with an 'A' profile, maximising the surface area and providing dense cover at the base. In conservation terms, the best solution is probably to have a mix of hedges, at various stages of growth and under various forms of management, within any area. Also to be recommended is the arrangement – encouraged by Countryside Stewardship schemes – of combining hedges with wide, unploughed headlands. The latter are not only of considerable conservation value in their own right – good for insects and an encouragement for the grey partridge – but they also protect the hedge from the harmful effects of herbicides and pesticides, sprayed on the adjacent fields. Nevertheless, care must still be taken to prevent fertilisers drifting onto these margins, especially if new hedges have been planted: tall, rank weed growth can suppress the development of the young shrubs.

In South Hams in Devon the National Trust is working in close partnership with the Royal Society for the Protection of Birds to boost the local population of cirl bunting. Sensitive management of the farmland and hedges is paying dividends, and numbers of this rare and delightful bird are increasing.

Oxborough, Norfolk. The hedge on the left was hacked back in early September, thus depriving local wildlife of a major source of food in the form of hips and haws. That on the right has, more sensibly, been left until later in the autumn before cutting.

Hedges, like trees, are increasingly thought of in positive terms, but that does not mean that they should be planted everywhere. Some landscapes have never contained hedges: coastal marshes, for example, where properties are separated by water-filled dykes; or areas of open heathland. Establishing hedges in such places destroys their essential, historic character: but there are more common, and more subtle, ways of eroding a distinctive sense of place. Most landowners prefer to plant hedges containing a wide range of shrubs, but in districts of late enclosure, characterised by species-poor hedges, this is not really appropriate. Hedges rich in hazel, dogwood and field maple are largely alien to landscapes like these. As they look out of character and out of place. Much better to plant a hedge of hawthorn or sloe, and reserve other species for creating new copses in inconvenient field corners. Arguably the best rule of thumb for planting is a simple one: copy the species found in neighbouring hedges, not least because these will be the plants best suited to the local soils and climate.

Field walls have not made quite the same comeback as hedges, and for obvious reasons. No one has yet devised a way of repairing walls by mechanical means, and it is much, much cheaper to plant a new hedge than to build a new wall. But while new drystone walls are a rare sight, existing ones are now treated with more respect. Over the last few decades their importance has become more widely recognised by the government, and by conservation bodies. The British Trust for Conservation Volunteers, and organisations like the National Trust, run regular working holidays centred around walling projects. Government grants for the repair and maintenance of walls have become available, especially in national parks like the Peak District. The craft of walling has itself experienced a welcome renaissance, partly as a result of the energetic activities of the Dry Stone Walling Association of Great Britain.

A vigorously sprouting hedge, newly rejuvenated by plashing, on the National Trust estate at Llanerchaeron in South Wales.

This limestone wall near North
Rauceby, like many in Lincolnshire,
serves no useful purpose in an
intensively arable county. Unloved
and unvalued, it is steadily decaying.

The overall attitude to field boundaries has clearly improved over the last two
decades, but there is certainly no room for complacency. Many landowners and
tenants still abuse their field boundaries. In particular, hedges continue to suffer
from the 'drift' of herbicide sprays, from intensive grazing, and from too regular,
and over-enthusiastic, flailing. Above all, they can deteriorate through simple,
sustained neglect, a policy which can convert an unwanted hawthorn hedge into
a line of disjointed and unstable trees. Lack of maintenance can equally signal
the gradual demise of a traditional drystone wall. And even though hedges are
reappearing in some districts, this is usually on roadsides, and on the margins
of existing fields, rather than as a result of the subdivision of vast prairies. The
arable east of England will remain a land of very large fields, even if these are
increasingly surrounded, once again, by hedges.

Some distinctive forms of boundary continue to be lost. Water-filled ditches are not
protected by any legislation, and their replacement by underground pipes, as
fields are amalgamated, continues in some low-lying districts. Dykes and ditches
have somehow never loomed quite so large in the minds of the public, or of
conservationists, as hedges and walls. Other kinds of boundary are simply
doomed in the long term. The pine rows of the East Anglian Breckland grow
tall and old: when wind or disease take their toll, few landowners make any
effort to replant, and so this evocative symbol of the local landscape will
eventually disappear.

Water-filled ditches, like this example on the Somerset Levels, have immense significance for wildlife but are afforded little protection by current legislation.

And here, perhaps, is a moment to pause and reflect. Those twisted trees were never intended to look like that. They were planted, for the most part, as low windbreaks and hedges. They are now redundant, neglected features of the landscape, and when they are replanted it is with the intention of ultimately recreating that picturesque and over-mature appearance, not of establishing a dense, managed hedge. Is this conservation or historical pastiche? Indeed, our new attitude to traditional field boundaries more generally tells us much about our wider relationships with the countryside, the past and the natural world.

We have seen how the character of hedges and walls reflects the times in which they were created: how, for example, early hedges were often planted with a range of species, so they could supply landowners with a number of necessary commodities. Today, the majority of walls and hedges are irrelevant to the needs of modern agriculture. A sceptic might argue that our new hedges, with their grow-tubes and mulch mats, improbably planted with hazel, dogwood, guelder rose and the rest in imitation of some ancient boundary, are as much an artefact of our escapist and heritage-conscious times as the no-nonsense, single-species, ruler-straight hedge was a symbol of the more practical world of the nineteenth century.

But I would take a more positive view. Our new attitude to field boundaries is a timely recognition that the countryside has more diverse functions than mere food production. Boundaries give us our sense of place, and provide a link with the past – both deeply-felt needs in an increasingly rootless age, as the success of organisations like Common Ground makes clear. Hedges, walls, banks and dykes provide the best way of allowing nature to thrive in a cultivated landscape. And they provide interest and variety both for local residents and for the thousands of visitors to country areas in this most crowded of islands. We should continue to protect, investigate, celebrate and enjoy this rich legacy.

Construction of a new drystone wall at the National Trust's Crickley Hill on the Cotswold escarpment. At over £100 per metre new walls are expensive; they are also prone to vandalism and theft, hence the setting of the 'toppers' into a mortar cap.

Further Reading

The following is a list of useful books and articles intended for those who wish to take the study of field boundaries further. Many can be obtained from public libraries; the more obscure can usually be ordered by them.

There is a vast literature on hedges and 'hedge-dating'. The best books to begin with are W. H. Dowdeswell, *Hedgerows and Verges* (London 1987); M. D. Hooper, E. Pollard, and N. W. Moore, *Hedges* (London 1974); M. D. Hooper (ed.) *Hedges and Local History* (London 1971); A. Brooks, *Hedging: a practical conservation handbook* (Wallingford 1975); R. and N. Muir, *Hedgerows* (London 1987); and T. A.Watt and G. P. Buckley (eds) *Hedgerow Management and Nature Conservation* (London 1994).

Useful local studies include: S. Addington, 'The Hedgerows of Tasburgh', *Norfolk Archaeology* 37 (1978), pp. 70-83; J. Hall, 'Hedgerows in West Yorkshire: the Hooper method examined', *Yorkshire Archaeological Journal* 54 (1982), pp. 103-9; D. R. Helliwell, 'The Distribution of Woodland Plant Species in Some Shropshire Hedgerows', *Journal of Biological Conservation* 7 (1975), pp. 61-72; G. Hewlett, 'Reconstructing a Historical Landscape from Field and Documentary Evidence: Otford in Kent', *Agricultural History Review* 21, pp. 94-110; J. Hunter, 'The Age of Hedgerows on a Bocking Estate', *Essex Archaeology and History* 24 (1993); J. Hunter, 'The Age of Cressing Field Boundaries', *Essex Archaeology and History* 28, 1997; T. Hussey, 'Hedgerow History', *Local Historian* 1987, pp.327-42; E. Pollard, 'Hedges, VII. Woodland Relic Hedges in Huntingdonshire and Peterborough', *Journal of Ecology* 61 (1973), pp. 343-52; and A. Willmot, 'The Woody Species of Hedge with Special Reference to Age in Church Broughton Parish, Derbyshire', *Journal of Ecology* 68 (1980), pp.269-86.

Important general articles, with good critiques of the 'Hooper hypothesis', include: R. Cameron, 'The Biology and History of Hedges: Exploring the Connections', *Biologist* 31 (1984), pp. 203-8; and W. Johnson, 'Hedges: A Review of Some Early Literature', *Local Historian* (1978), pp. 195-204. On locally distinctive hedgerows, see *National Research on Locally Distinctive Hedgerows* (Land Use Consultants/Countryside Agency, London 1999).

Useful information on hedge plants can also be found in R. Mabey, *Flora Britannica* (London 1998); and G. Grigson, *An Englishman's Flora* (London 1958).

In contrast to hedges, comparatively little has been published on the history of drystone walls. General accounts include: L. Garner, *Dry Stone Walls* (Aylesbury 1995); A. Raistrick, *Pennine Walls* (York 1946); R.Tunnell, *Building and Repairing Drystone Walls* (London 1982); and A. Brooks, *Dry Stone Walling: a practical conservation handbook* (Wallingford 1989). The best detailed studies remain R. Hodges, *Wall-to-Wall History: the story of Roystone Grange* (London 1991); but see also D. Spratt, 'Orthostatic Field Walls on the North Yorks Moors', *Yorkshire Archaeological Journal* 1998, pp.149-157.

On modern developments in the landscape, and their effects on nature conservation, see: W. Baird and J. Tarrant, *Hedgerow Destruction in Norfolk 1946-1970* (Norwich 1973); B. Green, *Countryside Conservation: landscape ecology, planning and management* (London 1996); R. Mabey, *The Common Ground* (London 1980); P. J. Perry, *British Farming in the Great Depression* (London 1974); M. Shoard, *The Theft of the Countryside* (London 1980).

Useful general books on fields and boundaries include: O. Rackham, *The History of the Countryside* (London 1986); C. Taylor, *Fields in the English Landscape* (London 1975); and S. Wade Martins, *Farms and Fields* (London 1995).

For prehistoric fields, and the thorny issue of 'relict field systems' surviving within the modern countryside, see: A. Fleming, *The Dartmoor Reaves* (London 1985); W. Rodwell, 'Relict Landscapes in Essex', in H. C. Bowen and P. J. Fowler (eds), *Early Land Allotment* (London 1987); T. Williamson, 'The Scole-Dickleburgh Field System Revisited', *Landscape History* 20 (1999), pp.19-28.

The best books on medieval fields are A. H. R. Baker and R. A. Butlin (eds) *Studies of Field Systems in the British Isles* (Cambridge 1973); D. Hall, *Medieval Fields* (Aylesbury 1982); D. Hall, *The Open Fields of Northamptonshire* (Northampton 1995). On the origins of open fields, see T. Brown and G. Foard, 'The Saxon Landscape: a regional perspective', in P. Everson and T. Williamson (eds) *The Archaeology of Landscape: studies presented to Christopher Taylor* (Manchester 1998), pp. 67-94; C. Lewis, P. Mitchell-Fox, P. and C. Dyer, *Village, Hamlet and Field: changing medieval settlements in Midland England* (Manchester 1997); and the various chapters in T. Rowley (ed.) *The Origins of Open-Field Agriculture* (London 1981).

The literature on enclosure, and the agricultural revolution, is substantial. On the timing and distribution of enclosure in general, see:
I. Williamson, 'Understanding Enclosure', *Landscapes* 1 (2000), pp. 56-79; J. R. Wordie, 'The Chronology of English Enclosure, 1500-1914', *Economic History Review* 36 (1983), 483-505; J. A.Yelling, *Common Field and Enclosure in England 1450-1850* (London 1977).

On changing farming systems generally, see
R. C. Allen, *Enclosure and the Yeoman: the Agricultural Development of the South Midlands 1450-1850* (Oxford 1982): E. Kerridge, *The Agricultural Revolution* (London 1967); M. Overton, *Agricultural Revolution in England: the transformation of the agrarian economy 1500-1850* (Cambridge 1996); and S. Wade Martins and T. Williamson, *Roots of Change: farming and the landscape in East Anglia 1700-1870* (Exeter 1999).

On the landscapes created by pre-parliamentary enclosure in the Midlands see: J. Broad, 'Alternate Husbandry and Permanent Pasture in the Midlands, 1650-1800', *Agricultural History Review* 28 (1990), pp.77-89; M. Reed, 'Pre-Parliamentary Enclosure in the East Midlands, 1550 to 1750, and its Impact on the Landscape', *Landscape History* 3 (1981), pp.60-68.

On parliamentary enclosure see: G. E. Mingay, *Parliamentary Enclosure in England: an introduction to its causes, incidence and impact* (London 1997); M. Turner, *English Parliamentary Enclosure* (Folkestone 1980); and M. Turner, *Enclosures in Britain 1750-1830* (Basingstoke 1984).

On the social effects of enclosure, the best books are J. M. Neeson, *Commoners; common right, enclosure and social change in England 1700-1820* (Cambridge 1993); K. D. M. Snell, *Annals of the Labouring Poor: Social change and Agrarian England 1660-1900* (Cambridge 1985). For the effects on the landscape see E. and R. Russell, *Making New Landscapes in Lincolnshire: the enclosure of 34 parishes* (Lincoln 1983); R. Carr, *English Fox Hunting: a history* (London 1976); A. Harris, *The Rural Landscape of the East Riding of Yorkshire, 1700-1850* (London 1961); J. Patten, 'Fox Coverts for the Squirearchy', *Country Life* 23 September 1971, pp.726-40; and M. Turner, 'The Landscape of Parliamentary Enclosure', in M. Reed (ed.) *Discovering Past Landscapes* (London 1984), pp.132-66.

The draining of the fens is dealt with in H. C. Darby, *The Changing Fenland* (Cambridge 1983); and in R. Hill's excellent *Machines, Mills and Uncountable Costly Necessities* (Norwich 1967).

Places to visit

The National Trust cares for almost 250,000 hectares of the most beautiful countryside in England, Wales and Northern Ireland, as well as for some 600 miles of superb coastline. Many of its properties include outstanding examples of field boundaries and offer excellent opportunities to both study these features at close quarters and to enjoy the landscapes in which they play such a key role.

The properties below are just a selection; fuller information on many more places to visit of landscape and wildlife interest is available in the *National Trust Coast & Countryside Handbook* or from the Trust's network of regional offices, details of which can be obtained from the following address:

> The National Trust Membership Department
> PO Box 39, Bromley, Kent, BR1 3XL
> tel. 0870 458 4000, fax 020 8466 6824
> enquiries@thenationaltrust.org.uk

County   Derbyshire

Property   The White Peak

The National Trust cares for various properties in the part of the Peak District known as the 'White Peak'. So-called because of its pale limestone walls, this area sits within a triangle bounded roughly by Buxton, Bakewell and Ashbourne. The high plateau is thronged with dramatic drystone walls, and many of the settlements are surrounded by interesting mosaics of ancient fields – Monyash and Wetton being particularly fine examples.

Access
A good starting-point is the National Trust car park and information centre at Ilam Country Park, a few miles north-west of Ashbourne. From here there are stunning views towards Thorpe Cloud and the entrance to Dovedale, and the park contains some well-preserved remains of medieval ridge and furrow. The whole White Peak area repays careful exploration, both by road and on foot.

When to visit
Although sometimes bleak, winter always provides an interesting perspective on this stunning landscape. The dramatic structure of the walls and field systems is at its most apparent at this time of year, and the views are often superb.

County   Cornwall

Property   The Lizard

The Lizard peninsula is one of the most dramatic places in England. It culminates in Lizard Point, the southernmost extremity of mainland Britain, and is famed for its outstanding views, complex geology and rare native plants. The area is laced with ancient and interesting field boundaries, with many of the banks faced in serpentine, a richly coloured and greasy-looking stone which was once quarried extensively and used for architectural features and domestic ornaments. The ruins of a nineteenth-century serpentine-cutting works can be seen at Carleon Cove, and today small workshops at Lizard Point and in Lizard village continue to produce serpentine ornaments.

Access

There is a car park in Lizard village, from where a footpath leads to Lizard Point. Several years ago the National Trust created one of the longest stretches of new Cornish 'hedge' along one side of this path: now maturing well, the hedge is full of wild flowers and invertebrate life. The coastal footpath leads from the Point along the eastern side of the Lizard to the villages of Cadgwith, Coverack and St Keverne, whilst to the west paths lead to delightful Kynance Cove.

When to visit

Spring and summer are ideal for enjoying the spectacular displays of wildflowers and interesting birdlife along the cliffs.

County   Cumbria

Property   Wasdale Head (Leconfield Commons)

A spectacular area of valley land and fell, with dramatic views of some of the highest peaks in the Lake District. There are many interesting historical features, including Bronze Age cairns, Iron Age and Roman settlements, and prehistoric field systems. The local grassland, heath and bogs are of great botanical importance, and the area is justly famed for its outstanding network of drystone walls.

Access

There is a National Trust car park at Wasdale Head (grid ref NY182074), from where footpaths lead through the surrounding ancient field systems. These paths provide excellent views of the area's stone walls and some eventually lead to Great Gable and Scafell Pike. An information vehicle is present on site during the main season.

When to visit

Autumn is particularly atmospheric: as the bracken dies off, it carpets the slopes in a deep russet and this, combined with the frequent autumnal mists, makes for quite magnificent scenery.

County   Dorset

Property   Golden Cap Estate

This large estate extends along eight miles or so of dramatic west Dorset coastline and includes Golden Cap, the highest cliff on England's south coast and so-called because of its exposed face of golden sandstone. The rolling countryside inland is characterised by small fields and flower-rich meadows bounded by ancient hedgerows. Some of these are more than four hundred years old and survive as fascinating evidence of the area's historical land use.

Access
There is a National Trust shop and car park at Stonebarrow Hill (off the A35 at Charmouth). More than twenty-five miles of footpaths and bridleways criss-cross the estate, giving excellent access to all the main habitats and interesting landscape features. The South West Coast Path runs along the clifftop.

When to visit
The ancient hedges are at their best in high summer, when the flowers and ferns they support are particularly abundant, and in autumn, when they are laden with berries and fruits.

County   Norfolk

Property   Horsey Estate

A remote and extensive area of wetland in the Norfolk Broads, containing a range of interesting habitats and some classic examples of drainage dykes. On one of these stands Horsey Windpump, a drainage mill built in 1912 but damaged in the 1940s by lightning and since restored by the National Trust. There has been a windpump on this site since the eighteenth century, used to regulate water levels and prevent flooding of the surrounding agricultural land. The wetland wildlife in the area is superb and includes a wide variety of birds, as well as water voles and swallowtail butterflies.

Access

A circular walk leads from the National Trust car park at the windpump (on the B1159) and gives excellent views of Horsey Mere and the adjacent fen and dykes. Information is available on site; the windpump is open seasonally and has a small shop and light refreshments. A further car park at Horsey Gap gives access to the beach.

When to visit

Outstanding throughout the year, but late spring provides the best opportunity to enjoy the large population of breeding birds along the dykes and around the mere. The swallowtails are at their best during June, with a second (and smaller) flight in August.

National Trust Publications

The National Trust publishes a wide range of books that promote both its work and the great variety of properties in its care. In addition to more than 350 guidebooks on individual places to visit, there are currently over 70 other titles in print, covering subjects as diverse as gardening, costume, dining and heraldry. These are all available via our website **www.nationaltrust.org.uk/bookshop** and through good bookshops worldwide, as well as in National Trust shops and by mail order on 01394 389950. The Trust also has an academic publishing programme, through which books are published on more specialised subjects such as specific conservation projects and the Trust's renowned collections of art.

Details of all National Trust publications are listed in our books catalogue, available from The National Trust, 36 Queen Anne's Gate, London SW1H 9AS – please enclose a stamped self-addressed envelope.

Hedges and Walls is the first title in a new National Trust series *Living Landscapes*. Appreciation of landscape dates back centuries, but a balanced understanding of the value of human interaction with the environment has only come about more recently. Through *Living Landscapes* we aim to explore this interaction, drawing on the vast range of habitats and landscapes in the Trust's care and on the bank of expertise the Trust has acquired in managing both these and the wildlife they support.

Each book in the series will explore the social and natural history of a different type of landscape or habitat: forthcoming titles will look at the fascinating world of *Parkland*, at the extraordinary story of our *Rivers and Canals*, and at the evocative subject of *Heathland*. Beautifully illustrated with specially commissioned artwork and a range of stunning contemporary photographs and historical material, this series will appeal to all those with an interest in social history, wildlife and the environment. Further details are available on our website (see above).

Picture credits

Opp. Contents	NTPL/Joe Cornish
pp8-9	Archie Miles
p10	NTPL/Derek Croucher
p11	Council for the Protection of Rural England
p12	Whitby Gazette
p13	NTPL/Geoff Morgan
p14	NT/Simon Ford
p15	Somerset County Council
pp16-7	Mary Evans Picture Library
p18	©Crown Copyright/NMR
p19	NT/Cornwall
p20	Tom Williamson
p20	NT/Chris Hill
p22	©Crown Copyright/NMR
p24	NTPL/Fay Godwin
p28	NTPL/Robert Morris
p30	NTPL/Joe Cornish
p33	NTPL/Robert Talbot
p34	Mary Evans Picture Library
p36	NT/Brenda Norrish
pp40-1	D. Woodfall/Woodfall Wild Images
p42	NTPL/John Hammond
p43	©Crown Copyright 1953. Reproduced with the permission of the Controller of Her Majesty's Stationery Office
p45	Mary Evans Picture Library
p46	Bedfordshire & Luton Archives & Records Office
p48	R.White/YDNPA
p50	©David Hosking/FLPA
p54	Hastings Museum & Art Gallery
p55	Reproduced from H. Saunders, *Illustrated Manual of British Birds*, 1899 edition
p58	Mary Evans Picture Library
p59	Reproduced from T. Williamson & L. Bellamy, *Property and Landscape*, 1983
p60	Mary Evans Picture Library (from Kerner-Leben, *Pflanzenleben*, 1913)
p62	D. Woodfall/Woodfall Wild Images
p63	D. Woodfall/Woodfall Wild Images
p64	Environmental Images/©Neil Lukas
p66	Council for the Protection of Rural England
p67	NT/Matthew Oates
pp68-9	Mary Evans Picture Library
p73	Rural History Centre, University of Reading
p74	Council for the Protection of Rural England
p75	NT/Nigel Hester
p76	NT/Angus Wainwright
p79	Archie Miles
p82	Tom Williamson
p83	James Parry
p85	Archie Miles
p86	Archie Miles
p88	James Parry
p89	(both) Archie Miles
p98-9	NTPL/Joe Cornish
p100	NTPL/Nick Meers
p101	Tom Williamson
p102	Tom Williamson
p103	Tom Williamson
p104	NTPL/Joe Cornish
p105	NTPL/Joe Cornish
p106	(both) Archie Miles
p107	NT/W. Hocking
p109	NTPL/John Darley
p110	(top) NT/Simon Ford
p110	(bottom) Tom Williamson
p117	NT/Giles Clotworthy
p118	NT/David Gowan
p120	NTPL/Joe Cornish
p121	Gwent Wildlife Trust
p123	Somerset County Council
pp128-9	NTPL/Joe Cornish
p130	NTPL/Ian Shaw
p132	David T. Grewcock/FLPA
p133	James Parry
p134	NTPL/Chris King
p135	Tom Williamson
p136	H. Clark/FLPA
p137	NTPL/Ian Shaw
p141	NTPL/Will Curwen
p143	NT/Jane Gifford
p145	NTPL/Joe Cornish
p147	NTPL/David Noton
p149	NTPL/Jonathan Cass
back cover	Paul Sterry/Nature Photographers Ltd

All inside artwork by Dan Cole
Maps by Artista

Index

C

D

E

F

G

H

X

Y

Z